研究生“十四五”规划精品系列教材

Quality Graduate Teaching Materials for the 14th Five-year Plan of Xi'an Jiaotong University

Fundamentals of tribology

摩擦学基础

曾群锋 主编

图书在版编目(CIP)数据

摩擦学基础=Fundamentals of tribology ：英文/曾群锋主编. -- 西安 ：西安交通大学出版社，2024. 10.
ISBN 978-7-5693-3841-6

Ⅰ. O313. 5

中国国家版本馆 CIP 数据核字第 2024YV8149 号

Fundamentals of tribology

摩擦学基础

主　　编　曾群锋
责任编辑　李　佳
责任校对　王　娜
装帧设计　伍　胜

出版发行　西安交通大学出版社
（西安市兴庆南路 1 号　邮政编码 710048）
网　　址　http://www.xjtupress.com
电　　话　(029)82668357　82667874(市场营销中心)
(029)82668315(总编办)
传　　真　(029)82668280
印　　刷　西安五星印刷有限公司

开　　本　787mm×1092mm　1/16　**印张**　9.625　**字数**　183 千字
版次印次　2024 年 10 月第 1 版　2024 年 10 月第 1 次印刷
书　　号　ISBN 978-7-5693-3841-6
定　　价　26.80 元

如发现印装质量问题，请与本社市场营销中心联系。
订购热线：(029)82665248　(029)82667874
投稿热线：(029)82668818

Preface

Tribology is a science, engineering and technology of interacting surfaces in relative motion under load and includes the study and application of the principles of friction, wear, and lubrication. Tribology as a subject is relatively new, but the practice of the tribological principles has a long history compared with the formal study of tribology. The tribological knowledge has been expanded tremendously in diverse fields in recent years, which leads to the demand for an improved understanding of tribology. Nowadays, tribology is an area of the active research in academic and industrial fields. Although there is a large amount of literature available on this subject to the researchers, there is a paucity of literature for engineers and in college courses due to the complexity and interdisciplinary nature of the subject. While offering course on tribology, we find it difficult to recommend a textbook that covers all the aspects of tribology for the students. The present book is expected as an introductory comprehensive textbook about the field of tribology. Thus, this book aims to collect all the bits of knowledge including new research achievements in tribology. This book consists of 11 chapters, and begins with an introduction to tribology. The history, industrial application, origin, and importance of tribology are also well-discussed. Then it is followed by Chapter 2 on engineering surface and interface. Chapter 3 deals with friction. Wear is the subject matter of Chapter 4. Chapter 5 concerns lubrication theory. Superlubicity is outlined in Chapter 6. Bionics tribology is the subject matter of Chapter 7. Chapter 8 addresses green tribology. Nanotribology is listed in Chapter 9. Tribology under extreme environments is discussed in Chapter 10. Finally, the tribology of basic components is listed in Chapter 11. Each chapter contains a summary and the perspective for future study. Although this book can be used both as a textbook and a reference book, one of its major objectives is to highlight many recent developments in the field of tribology. The book targets the undergraduate and postgraduate students, practising engineers, and researchers in various academic institutions. It is expected that this book will be a valuable guide and a useful reference for those who are working in the area of tribology. Many thanks are dedicated to all my graduate students who have spent considerable time in the preparation of various chapters.

It can also serve as a reference book for engineers and designers. I am grateful to all my colleagues and graduate students for their contribution to this book.

I would appreciate comments and suggestions from the readers for the improvement of the textbook.

Zeng Qunfeng
March, 2024

Contents

1 Introduction to Tribology ······ 1
1.1 History of Tribology ······ 1
1.2 Industrial Applications of Tribology ······ 4
1.3 Origin of Friction ······ 6
1.4 Importance of Tribology ······ 8
1.5 Layout of the Book ······ 8
1.6 Brief Summary ······ 8
2 Engineering Surface and Interface ······ 9
2.1 Nature of Surface ······ 9
2.2 Physico-Chemical Characteristics of Surface ······ 11
2.3 Definition of Surface Roughness ······ 12
2.4 Measurement of Surface Roughness ······ 14
2.5 Surface Tension ······ 15
2.6 Surface Contact ······ 16
2.7 Brief Summary ······ 19
3 Friction ······ 20
3.1 Introduction ······ 20
3.2 Sliding Friction ······ 21
3.3 Rolling Friction ······ 22
3.4 Friction Theory ······ 23
3.5 Friction Model and System ······ 26
3.6 Fluid Friction ······ 27
3.7 Friction of Materials ······ 28
3.8 Brief Summary ······ 29
4 Wear ······ 30
4.1 Introduction ······ 30
4.2 Types of Wear ······ 31
4.3 Prediction of Wear ······ 33
4.4 Wear of Materials ······ 35
4.5 Wear Control ······ 36
4.6 Wear Under Specific Conditions ······ 38
4.7 Brief Summary ······ 40
5 Lubrication Theory ······ 41

5.1 Introduction …… 41
5.2 Stribeck Curve and Lubrication Status …… 41
5.3 Boundary Lubrication …… 43
5.4 Fluid Lubrication …… 46
5.5 Lubricants …… 49
5.6 Brief Summary …… 51
6 Superlubricity …… 52
6.1 Introduction …… 52
6.2 Liquid Superlubricity …… 53
6.3 Solid Superlubricity …… 57
6.4 High Temperature Superlubricity …… 60
6.5 Super Low Wear …… 61
6.6 Superlubricity Mechanism …… 63
6.7 Superlubricity System …… 65
6.8 Brief Summary …… 66
7 Bionics Tribology …… 67
7.1 Introduction …… 67
7.2 Definition and Fundaments …… 67
7.3 Bio-tribology …… 69
7.4 Development of Friction Elements …… 74
7.5 Manufacture of Bionic Tribology Surface …… 75
7.6 Bionic Lubrication and Lubrication Systems …… 77
7.7 Tribological Properties of the Bionic Tribology System …… 78
7.8 Applications and Developments …… 80
7.9 Brief Summary …… 81
8 Green Tribology …… 82
8.1 Introduction …… 82
8.2 Green Tribology …… 83
8.3 Eco-Friendly Lubrication …… 87
8.4 Green Lubricants and Materials …… 91
8.5 Smart Lubrication …… 93
8.6 Sustainable Tribology …… 95
8.7 Brief Summary …… 96
9 Nanotribology …… 97
9.1 Introduction …… 97
9.2 Nano Friction …… 97
9.3 Nano Wear …… 101
9.4 Nano Lubrication …… 103
9.5 Tribochemistry in Nanotribology …… 104

9.6 Atomic-Scale Computer Simulations …… 106
9.7 Application and Development of Nanotribology …… 108
9.8 Brief Summary …… 110
10 Tribology Under Extreme Environments …… 111
10.1 Introduction …… 111
10.2 High Vacume Tribology …… 112
10.3 Tribology at Extreme Temperatures …… 116
10.4 High Speed Tribology …… 119
10.5 Tribology in the Ocean Environment …… 121
10.6 Multi-Field Coupling Tribology …… 123
10.7 Brief Summary …… 124
11 Tribology of Basic Components …… 125
11.1 Introduction …… 125
11.2 Journal and Thrust Bearing …… 125
11.3 Rolling Bearing …… 126
11.4 Gear …… 128
11.5 Mechanical Seal …… 132
11.6 MEMS …… 134
11.7 Rail Transport Tribology …… 137
11.8 Reliability of Tribology …… 140
11.9 Brief Summary …… 144
References …… 145

1 Introduction to Tribology

Tribology was first used by Peter Jost in 1966. Tribology is the theory and technology to study the friction, wear, and lubrication between the interacting surfaces with relative motion or relative motion trends, and the interrelationships among them. Tribology involves mathematics, physics, chemistry, material science, metallurgy, mechanics, mechanical engineering, chemical engineering, and other disciplines.

1.1 History of Tribology

Tribology has been in existence since the beginning of the recorded history.

1.1.1 Development of Tribology

Tribology has a continuous evolution. The tribological developments are traced back to the prehistoric epoch (Stone Age) until 3500 before Christ (B.C.), when the generation of fire by friction between two pieces of wood is considered as one of the first examples of the tribological applications. In Egypt, and dated around 2000 B. C. depictions show the transport of stone colossuses on slides, men poured liquids (lubricants) appearing in front of the slides. The Greek and Roman epoch, from 900 B. C. to 400 A.D., is characterized by the developments in journal bearings and gears. A famous example is the rolling bearing for rotating platforms on Roman ships found at the bottom of Lake Nemi in Italy, dated around 50 A.D. Vegetable and animal oils are used as lubricants. Metallic bearings in China were developed in the year of 900 A.D. and several mechanical clocks were also built with metal gears and brass journal bearings. Leonardo da Vinci, who first studied friction phenomena, proposed in 1470 that the friction force is proportional to the load, but independent of the nominal contact area. He deduced the laws governing motion and introduced the concept of the coefficient of friction (CoF), the ratio of friction force to the normal load. Amontons proposed in 1699 that friction is caused by work, and published two laws of friction. Coulomb put forward the third law of friction in 1785. The laws of friction are listed as follows:

(1)The friction force is proportional to the normal load.

(2)The friction force is independent of the nominal contact area.

(3)The friction force is independent of the sliding speed.

From 1750 to 1850, gears, journal bearings, roller bearings, and lubricants were developed. The viscosity was coined by Navier and defined by Stokes in the well-known equations. Poiseuille performed the studies on fluid flow in a pipe. Stribeck determined the curve showing the relationship between friction, load, speed, and viscosity. Sommerfeld proposed an analytical solution to the Reynolds' equation and the Sommerfeld number. Reynolds established the basic equation of hydrodynamic lubrication in 1886, and proved that the hydrodynamic pressure of liquid entrained between two sliding surfaces is sufficient to prevent contact between two surfaces. Bowden and Tabor proposed that friction is caused by the adhesion between micro-peaks and established the theory of adhesion in 1950. In 1953, Archard formulated the law of wear, stating that the amount of wear is proportional to the normal force and the sliding distance, and inversely proportional to the hardness of the material. In 1957, Khruschov proposed the abrasive wear resistance equation of metals. In 1959, Dowson predicted the minimum and central film thickness values of elastohydrodynamic lubrication (EHL). In 1962, Johnson proposed a shakedown limit in rolling contact. Jost presented the Tribological Report (Jost Report) in 1965. Recently, new areas in tribology have emerged, including nanotribology, biotribology, and superlubricity. Hirano proposed the concept of superlubricity to describe a theoretical sliding regime where friction between two contact surfaces almost vanishes in 1990. Theoretically, superlubricity is the realization of zero friction force. But in practice, it is considered as the situation when CoF is less than 0.01. Klein reported liquid superlubricity whose CoF in the order of 0.001 in low-density polymer brushes in 1991. Martin reported solid superlubricity of MoS_2 coatings in ultrahigh vacuum in 1994. In 2001, Xie proposed three axioms of tribology as follows:

(1) Tribological behaviors are system-dependent.

(2) The tribological of tribo-elements is time-dependent.

(3) The results of the tribological behaviors are the results of the simultaneous acting and strong coupling of many processes of multi-disciplines.

In 2010, Zeng achieved a high-temperature solid superlubricity of about 0.008 at 600 ℃. In 2013, Zeng proposed the combined solid-liquid superlubricity of diamond-like carbon (DLC) films under Polyalphaolefin (PAO) oil with nano boron nitride additive. In 2015, Erdemir realized the macroscopic superlubricity through the incommensurate contact between DLC films and the graphene nanoscroll. In 2024, Aymard proposed a surface design strategy to prepare dry rough interfaces that have predefined relationships between normal and friction forces. With the development of surface science, material science, and artificial intelligence (AI), tribology has developed rapidly in recent years. Fig. 1-1 shows a brief development of tribology.

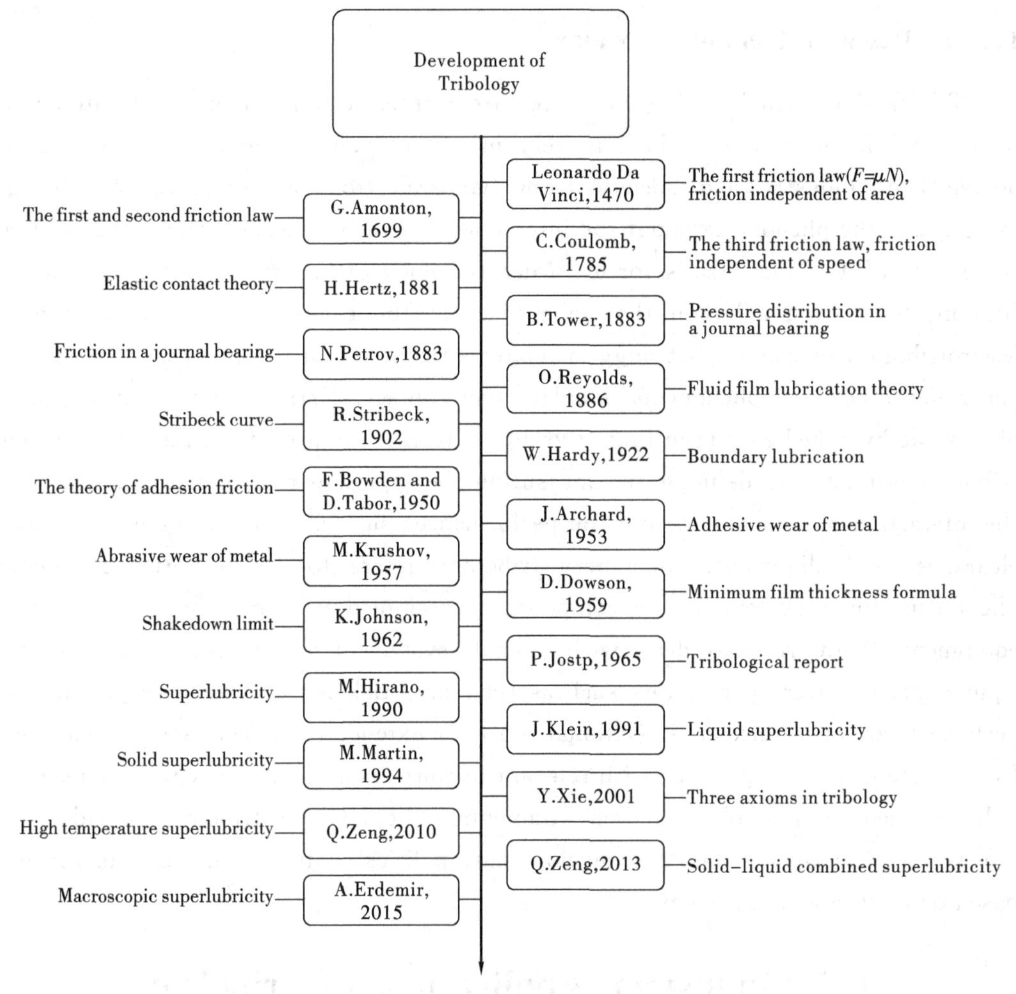

Fig. 1 - 1 Development of tribology

1.1.2 Fundamentals of Tribology

Tribology mainly includes friction, wear, and lubrication. These areas make up the fundamental aspects of tribology. Friction is defined as the resistance to the movement of a body against another. Friction is not a material parameter, but a system response in the form of a reaction force. The friction force is divided into two components, namely, the adhesive component and the ploughing component, respectively. Wear including abrasive, adhesive, fatigue, and corrosion wear occurs due to the interaction between two surfaces in contact and implies the gradual removal of the materials of the surface. Wear of materials in contact is a system parameter. Low friction usually results in low wear. However, it is not a general rule and numerous examples have shown high wear rates despite low friction. Lubrication, which is the process or technique of using a lubricant, plays a crucial role in reducing wear by minimizing friction. Lubricants are primarily used to reduce friction and minimize wear between the interacting surfaces.

1.1.3 Research Field of Tribology

The field of tribology has many features: interfacial phenomena of interacting bodies in relative motion, which is very important and responds to the increasing demands of industry and society from nanoscale to macroscale. Nanotribology investigates the phenomena about the interaction between molecules and atoms such as the effects of Van der Waal's forces. Microtribology or asperity tribology studies of friction, wear, and adhesion that take place at the peaks of surface topography. Macrotribology or contact tribology is related to contacts in gears, bearing elements, and rollers, besides, phenomena like Hertzian contact, EHL, and wear mechanisms observable by naked eyes (scuffing, scoring, pitting) are also of interest. Component tribology is related to defining and measuring the typical parameters originating from the interaction of components and the performances such as torque, force, vibration, clearance, and alignment. Machinery tribology refers to the performance related phenomena for a system of the components assembled in a machine or a piece of equipment. Plant tribology deals with a whole system of machinery, structures, and equipment; however, parameters such as economy, risk levels, availability, and life-cycle costs are also considered. National tribology extends the effects and consequences from a nationwide perspective, with relevant parameters such as safety policy, research policy, transportation policy, and environmental policy. Global tribology considers the effects that deal with sustainable development, politics, cultural, and human survival based on an interacting system.

1.2 Industrial Applications of Tribology

Tribological studies of machine components with relative motions in industrial applications are the focus of attention. These machine components include bearings, seals, gears, metal cutting, magnetic storage devices, micro-electro-mechanical systems (MEMS) as well as biomedical products.

1.2.1 Machine Components

The machine components involved in tribology include bearings, mechanical seals, cams, gears, and brakes.

Bearing is a kind of machine part that is used to support shafts and the load transmitted from the shaft. There is traction and wear between two surfaces such as the rolling between the rollers and the raceway of the inner or outer ring during operation. Cams and followers in internal combustion engines achieve the desired functions by making continuous contact between them. There are various failure mechanisms: the accelerated wear, scuffing, pitting, and polishing. Gears transfer power from one shaft to another in a wide range of applications. The gear failures include friction and scuffing, which involve scratching and rusting. Mechanical face seals are used in rotating equipment such as pumps, mixers, blowers, and compressors. The main components of a mechanical face seal are two rings, the stator and the rotor, which are arranged perpendicular to the axis of the rotating shaft. Disc brake systems are widely used in cars and commercial vehicles for braking. During braking, the rotor slides against the friction materials, transforming the kinetic energy of the vehicle into frictional heat.

1.2.2 Industrial Revolution and Industrial 5.0

The technology-driven industrial revolution conceptualizes the rapid changes in technology, industry, social patterns, and processes in the past few decades. Fig. 1 - 2 shows an overview of the evolution of industry. In the 1800s, Industry 1.0, the "Age of Steam", evolved through the development of mechanical production infrastructures for water and steam-powered machines so that the steam engine, the textile industry, and the mechanical engineering are the fundamental aspects of this age. The year of 1784, when the first power loom was built, is often considered as a key to this revolution. Industry 2.0, the "Age of Electricity", evolved in 1870, when the first assembly line was used on a large scale in slaughterhouses, and primarily focused on mass production and the distribution of workloads. Industry 3.0, the "Information Age", evolved in 1969 with the concept of electronics, partial automation, and information technology. Industry 4.0, the "Age of Cyber Physical Systems", evolved in 2011 with the concept of smart manufacturing for the future to maximize productivity and achieve mass production. Industry 5.0, it is regarded as the next industrial evolution to leverage the creativity of human experts in collaboration with efficient, intelligent, and accurate machines and obtain resource-efficient and user-preferred manufacturing solutions.

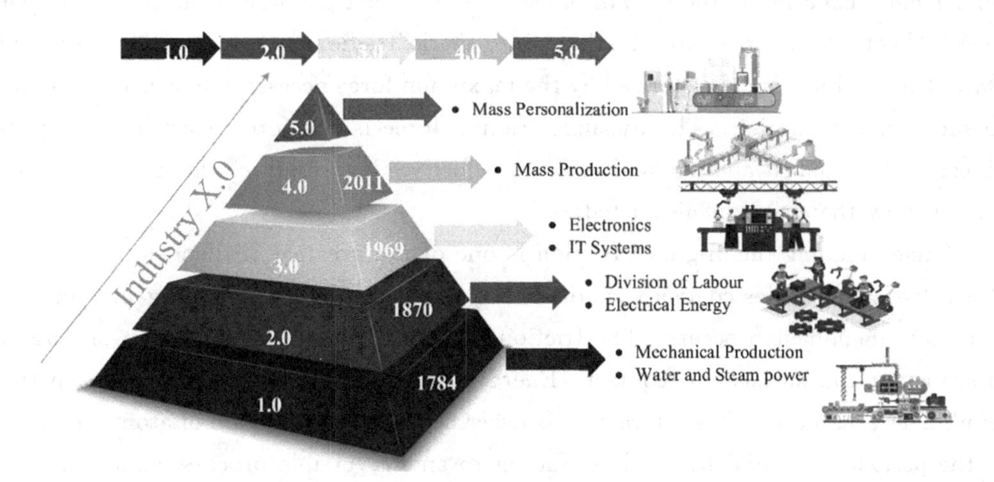

Fig. 1 - 2 Illustration of industrial evolution

1.2.3 Tribology Evolution

Tribology has had a continuous evolution over time. Tribology involves improvements in life prediction of components, investigation of failure mechanisms, diagnostic analysis and prognostic predictions, and maintenance of operating machinery. The efficiency, performance, maintainability, and service life of machine elements such as seals, bearings, and gears can be improved by the appropriate tribological design. Tribological studies are useful for the reduction of emissions from tyres and brakes. Novel and environmentaly-friendly lubrications and additives are under development. Renewable lubricants, water-, nitrogen, or hydrogen lubrication, anti-wear coatings, surface texturing, surface modification, non-metallic bulk materials, and life cycle tribology are indicated as the main future research developments of industrial tribology. Tribology in manufacturing is another important tribological research field. Tribology can contribute to the development of additive manufacturing with the tribological investigations on new printed products and investigate the surface mechanical characteristics.

1.3 Origin of Friction

Friction is a force that opposes the relative motion or tendency of motion between two objects. Friction is a complex phenomenon that involves various factors, such as adhesion, intermolecular forces, surface roughness, and molecular bonding. These forces may be attractive or repulsive, depending on the nature of the involved substances.

1.3.1 Nature of Friction

Friction plays a vital role in various fields of study, including mechanics, engineering, physics, and mathematics. The friction force is one of the most fundamental concepts in the field of physics, and it has a significant impact on our daily lives. The friction forces are classified into static friction and dynamic friction. The static friction force is characterized by the maximum force necessary to cause one subject to slide against another. The dynamic friction force is the force exerted to keep one object sliding against another with a finite velocity. The origin of the dynamic friction force is how the energy is dissipated.

Understanding the origin of friction is one of the greatest challenges in tribology. When two solid bodies contact each other and one body begins to slide against another, a friction phenomenon occurs. The friction force acts as resistance against sliding and arises along the surfaces in contact. Energy loss due to friction occurs as asperities climb over each other. The interaction is described as the interaction of atoms connected to the periodic potential field, thus leading to an irreversible process where energy is consumed as friction works. The friction force may originate from the shear and plastic deformation of the adhesive junctions, which are proportional to the actual contact area. The friction experiments between two molecular smooth non-adhering surfaces have been conducted to examine Amontons' law at the molecular scale and revealed that even at the microscopic level, the friction force is remained proportional to the net applied load.

1.3.2 Origin of Superlubricity

The tribologists have been trying to verify the possibility of achieving an absolute friction-free state and investigate the friction mechanism and the development of superlubricity, which has been demonstrated as an appealing way to achieve ultralow friction and wear with almost no energy dissipation.

Theoretically, superlubricity is the realization of zero friction force. But in practice, CoF remains typically below 0.01 due to measurement precision limitations and other factors. Superlubricity research is categorized into two areas. One is in theory, where most researches have focused on investigating the conditions of superlubricity and mechanisms of superlubricity. Another is the experimental area, where great efforts have been made to discover kinds of superlubricity materials. In addition, simulation also provides an effective tool for studying the energy dissipation in superlubricity. It is anticipated that the future application of superlubricity could yield significant economic benefits and offer substantial energy-saving potential applications.

1.3.3 Application of Tribology

Tribology plays a fundamental role in technological and economical fields. Industrial components or systems are significantly influenced by machine wear and friction losses. Energy conservation is a broad domain where tribology can notably impact the industrial imperative to minimize losses and wastes. Monitoring with the wear, friction, vibrations and temperatures provides data to enhance the performance of the tribological industrial systems. Tribology networks and databases containing tribological data can deliver important information through internet and wireless connections among several electronic devices.

The tribological performance of any systematic nano-micro or macro-scale depends upon a large member of external parameters such as temperature, contact pressure, and relative speed. Tribology is very imperative in various applications, including various kinds of bearings, gears, engines, orthopaedic joints and micro-machines. Detailed insights into these applications are provided in the following researche: bio-tribology, micro/nano tribology, wind turbines, nano-lubricants, automotive tribology, bearings, green tribology, and more.

1.4 Importance of Tribology

Tribology has a profound impact on many areas of engineering and everyday life. The interaction surface in relative motion is accompanied by tribology involving friction, wear, and lubrication. Friction between two surfaces causes energy to dissipate, resulting in loss of resources. The improvements in friction reduction technologies could reduce significantly friction energy losses in machinery parts. Reducing wear can enhance long-term efficiency and the performance of moving components, while also reducing maintenance costs and/or improving quality of life. Tribological studies of various components in industrial applications have been a focal point of attention for many years.

1.5 Layout of the Book

The present book aims to bulid the tribology community of engineers, material scientists, applied physicists, and chemists who work to solve the tribological problems and also aims to provide a starting point for further collaboration and possible focal points for future interdisciplinary research in tribology. Accordingly, the book is organized as follows: the basic concepts of tribology, research themes including superlubricity, bionic tribology, green tribology, nanotribology, and the industrial application of tribology.

Chapter 1 presents the introduction, definition, history, and development of tribology, along with industrial applications, origins and the significance of superlubricity. Chapter 2 discusses engineering surface and interface, surface roughness, surface tension and contact. Chapter 3 addresses sliding friction, rolling friction, theory, model, and friction materials. Chapter 4 covers the types and prediction of wear, wear of materials, and wear control. Chapter 5 explores lubrication, the Stribeck curve, boundary lubrication, fluid lubrication, and lubricants. Chapter 6 mainly talks about superlubricity. Chapter 7 presents bionic tribology, biotribology, bionic lubrication system, manufacture of bionic surface, and the applications of bionic tribology. Chapter 8 focuses on green tribology, eco-friendly lubrication, green lubricants, and sustainable tribology. Chapter 9 examines nanotribology. Chapter 10 introduces tribology under extreme condtions. Chapter 11 discusses the reliability of tribology.

1.6 Brief Summary

The present chapter introduces the theory, history, and development of tribology, from the traditional friction to the developed newly superlubricity, and so on. In addition, the industrial applications of tribology are also introduced.

2 Engineering Surface and Interface

The boundary between solid and gas phases is known as a surface, while the boundary between solid phases is referred to as an interface. Solid surfaces possess the complex structures, and their properties are influenced by various factors, including the inherent nature of the solid, the surface preparation method employed, and interactions with the surrounding environments. These surface properties play a crucial role in determining surface interactions such as real contact areas, friction, wear, lubrication, chemical reactivity, surface tension, and surface free energy.

2.1 Nature of Surface

Above the bulk material, there is usually a work-hardened layer resulting from the machining process. The outer part is often contaminated by materials from cutting or forming tools and the associated lubricants, and the superficial layer consists of the reaction products formed by chemical reaction with the environment, typically oxides, together with the contaminants such as dust and grease from the atmosphere. A representation of a rough and contaminated surface is shown in Fig. 2 - 1.

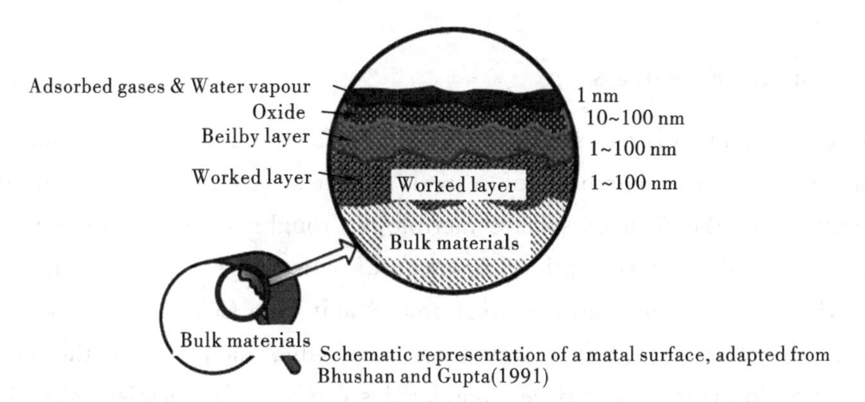

Fig. 2 - 1 A representation of a rough and contaminated surface

2.1.1 Physical Properties

Surface properties encompass surface energy, surface adsorption, and surface diffusion. Surface free energy, also known as surface tension, is described as the additional energy possessed by surface atoms. Overcoming surface energy is a key challenge in achieving super-low friction.

Adsorption is a surface phenomenon that involves the transfer of molecules from a fluid bulk to a solid surface. This process results from either physical forces or chemical bonds and is typically reversible. Adsorption on a solid surface occurs when molecules or atoms on the surface possess residual surface energy due to unbalanced forces. The adsorption process is classified into two categories: physical adsorption and chemical adsorption.

Physical adsorption is driven by intermolecular forces, such as Van der Waals forces. It generally occurs at low temperature, exhibits a fast adsorption rate, low adsorption heat, and is non-selective.

Chemical adsorption involves the formation and breaking of chemical bonds. Physical adsorption and chemical adsorption are often intertwined and coexist in surface process. Fig. 2-2 illustrates the comparison between physical adsorption and chemical adsorption.

Surface diffusion refers to the movement of the adsorbed molecules on surfaces.

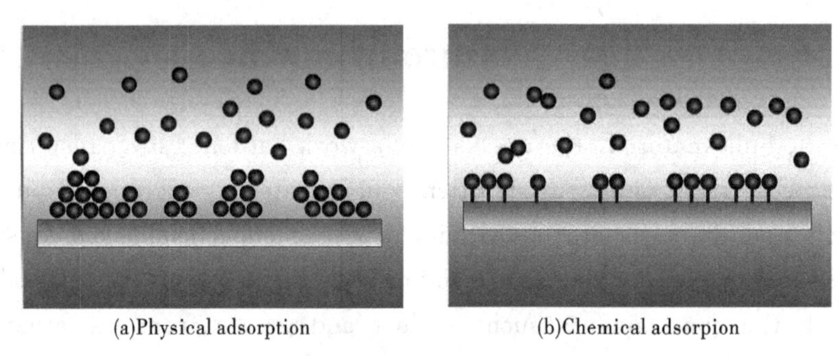

(a)Physical adsorption (b)Chemical adsorpion

Fig. 2-2 Physical adsorption and chemical adsorption

2.1.2 Surface Roughness

Surface roughness is an essential aspect of surface texture, which is measured by the disparities in the alignment of a real surface's normal vector from its ideal configuration. In the field of surface metrology, roughness commonly refers to the high-frequency, short-wavelength component of a measured surface. In tribology, rough surfaces tend to undergo fast wear and exhibit high CoF compared with smooth surfaces. Therefore, roughness serves as a valuable indicator of the mechanical component performance, as surface irregularities can serve as incipient sites for crack formation or corrosion.

2.1.3 Contact Surface

When two objects come into contact and exert pressure on each other, they generate a force on the contact surface known as friction. Friction acts in the opposite direction to the relative motion or the tendency of the relative motion between the objects. The friction between solid surfaces is attributed to two main factors. Firstly, it arises from the mutual attraction between atoms and molecules on solid surfaces.

Secondly, it is influenced by surface roughness in contact. Friction is influenced by surface roughness and the applied pressure.

2.1.4 Smart Surface

Smart surfaces are capable of self-assembling into a monolayer or polymer brush on a substrate's surface. These surfaces exhibit unique properties such as hydrophilicity, hydrophobicity, conductivity/insulation, and adhesion, which can be modulated through environmental stimuli. The changes include knotting, repulsion, adsorption, desorption, and other reversible transformations. The categories include solvent response, temperature sensitivity, electric field response, pH response, and photosensitive smart surfaces. This direction of research has great potential for the future development of smart surfaces.

2.2 Physico-Chemical Characteristics of Surface

A complete surface is composed of several distinct layers: a deformed layer, a chemically reactive layer, a physisorbed layer, and a chemisorbed layer.

2.2.1 Deformed Layer

The deformed layer is formed during the material preparation process, such as lapping, grinding, or polishing. It can also occur during the friction process. These surface layers undergo plastic deformation either with or without a change in temperature, leading to high-strain regions. The formation of the deformed layer primarily depends on the energy input during its creation and the properties of the involved material.

The thickness of the deformed layers typically ranges from 1 to 10 μm and from 10 to 100 μm, respectively. The mechanical properties of the surface are influenced by the extent and depth of deformation in this layer.

2.2.2 Physisorbed Layer

The physisorbed layer is formed through the interaction of the surface with molecules such as oxygen, hydrocarbons from the environment, and water vapours. These molecules undergo physical adsorption, leading to the formation of a layer. The molecular interaction occurs through weak Van der Waals forces, which may be easily removed when the surface interacts with other substances.

The physisorption process primarily involves Van der Waals forces. Occasionally, there may be a greasy or oily film, which can partially displace the adsorbed layer originating from the environment. The thickness of the greasy films can be as small as 3 nm.

2.2.3 Chemisorbed Layer

The chemisorbed layer is formed through the strong bonding interaction between molecules, such as covalent bonds. These chemical bonds are strong, requiring high energy for their removal. The removal of these layers often releases heat due to the strength of the chemical bonds involved.

2.2.4 Chemically Reactive Layer

Chemically reactive layers are formed as a result of the surface materials' reaction with atmospheric oxygen, leading to the formation of oxide layers. Other elements, like nitrides, sulfites, chlorides, and so on, may also react with the surface materials, leading to the formation of the respective oxide layers. These layers are predominantly formed during machining or friction processes.

2.2.5 Surface Characterization

Surface characterization involves a wide range of analytical techniques. To possess the metallurgical properties of a deformed layer, the surface is sectioned, and its cross-section is examined using a scanning electron microscope (SEM). The microcrystalline structure and dislocation density may be investigated with a transmission electron microscope (TEM). Additionally, the crystalline structure of a surface layer is analyzed through X-ray, high-energy or low-energy electron diffraction techniques. For chemical analysis, X-ray photoelectron spectroscopy (XPS) and secondary ion mass spectrometry (SIMS) are commonly used. The thickness of the layers can be accurately measured by conducting depth profiling of the surface. Furthermore, the chemical analysis of the adsorbed organic layers can be carried out using various surface analytical tools, including Fourier transform infrared spectroscopy (FTIR), Raman scattering, nuclear magnetic resonance (NMR) and XPS.

2.3 Definition of Surface Roughness

Surface roughness is the intricate landscape of a material's surface, characterized by microscopic valleys and peaks of varying sizes and dimensions. These features, with unique spacing and heights, contribute to the texture and quality of the surface, influencing functionality and performance in various applications.

2.3.1 Average Roughness Parameters

Surface roughness evaluation is very important for addressing many fundamental issues such as friction, contact deformation, heat and electric current conduction, tightness of contact joints and positional accuracy.

The amplitude parameter is the most important parameter for characterizing surface topography, as it measures the vertical character of surface deviations.

The spacing parameter measures the horizontal character of surface deviation. One of the spacing parameters is peak spacing, which is an important factor in the performance of friction surfaces such as a brake drum.

1. Center Line Average (CLA)

The arithmetic average height parameter, known as the Centre Line Average (CLA), is the most widely used roughness parameter for general quality control purpose. It is defined as the average absolute deviation of the roughness irregularities from the mean line over one sampling length. Fig. 2 - 3 provides a brief overview of surface roughness and related concepts.

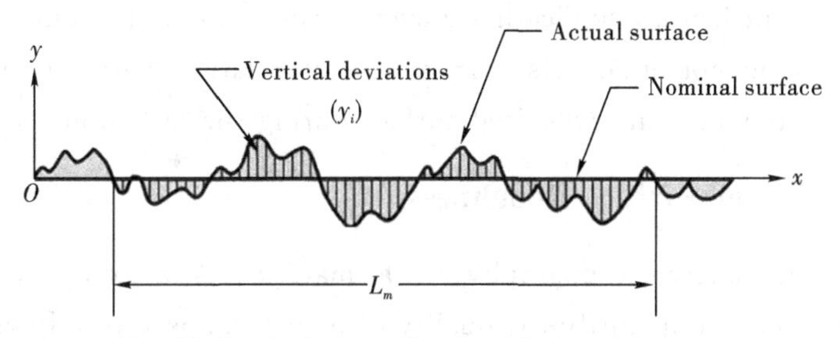

Fig. 2 - 3 Surface roughness and related concepts

2. Root mean square (RMS) roughness

Sq represents the root mean square value of the ordinate values within the definition area. It is equivalent to the standard deviation of heights.

3. Arithmetic mean surface roughness (R_a)

To calculate R_a, set a length L on a surface contour curve and take the center of the length as the X-axis. Divide the sum of the areas of all oblique lines within the length by the measured length L is the R_a.

4. Root mean square surface roughness (R_y)

Set the length L on the surface contour curve. The vertical distance from the highest peak to the lowest trough of the curve within this length is the maximum rough value, denoted as R_{max} or R_y.

5. Maximum surface roughness (R_z)

Set a length L on the surface contour curve, and measure the distance between the top of the fifth peak and the bottom of the fifth trough at the center of the curve within this length, denoted as R_z.

2.3.2 Statistical Analysis

The amplitude distribution function is a probability function that provides the likelihood of a roughness profile having a specific height at a given position. Skewness reveals the asymmetry of the amplitude distribution function. Positive skewness suggests a concentration of material near the base of the profile, while negative skewness suggests a concentration near the top. Kurtosis, on the other hand, measures the sharpness of the amplitude distribution function.

2.3.3 Fractal Analysis

Fractal theory is a branch of nonlinear science, which on the study of the complex and random geometric objects exhibiting self-similarity, affine transformations, and scale invariance. This theory finds extensive applications in natural sciences and engineering disciplines. In the field of surface engineering, self-similarity refers to the property of remaining invariant at different scales after equal transformations in all directions, while self-affinity describes the invariance of a surface when it is scaled by direction-independent factors, indicating unequal transformations from the global to the local level. This phenomenon arises from the correlation between newly adsorbed atomic layers and the adjacent deposited surface during the diffusion process.

2.3.4 Modelling of Surface Roughness

Predicting the surface topography of a material is a complex task. Surface roughness, often used to quantify the quality of a surface, is typically characterized by R_a. Analytical models for surface roughness have been developed based on the microstructure of the grinding wheel, considering both one-dimensional and two-dimensional aspects. These models simplify the wheel microstructure by assuming a constant distance between cutting edges and a uniform height for the cutting edges.

2.4 Measurement of Surface Roughness

The measurement techniques for surface roughness are categorized into two main types: contact and non-contact methods.

Contact-type measurement techniques involve physical contact between the measuring instrument and the surface being evaluated, which are often precise and can provide accurate measurements but they may be limited by factors such as probe size and potential surface damage.

Non-contact measurement techniques do not require direct physical contact with the surface. These methods utilize optical, laser, or profilometry technologies to capture the surface profile without touching it.

2.4.1 Optical Method

Optical interferometry is a highly valuable technique for precise surface shape measurement, which involves measuring surface heights at multiple visible wavelengths, effectively creating a synthetic wavelength that is significantly longer than the visible wavelengths used.

White-light scanning interferometry focuses on measuring the degree of fringe modulation or coherence rather than the phase of the interference fringes. By analyzing the fringe modulation, the surface shape and roughness can be accurately determined.

2.4.2 Scanning Probe Microscopy Method

Scanning probe microscope (SPM) is a versatile tool used for imaging nanoscale surfaces and structures, including individual atoms. It consists of a probe tip mounted on a cantilever, which may be incredibly sharp, even at the atomic level. The tip may be precisely and accurately moved back and forth across the surface, enabling atom-by-atom manipulation. When the tip approaches the sample surface, the cantilever experiences a deflection due to various forces, including mechanical contact, electrostatic forces, magnetic forces, chemical bonding, Van der Waals forces, and capillary forces. To measure the deflection, a laser is directed onto the top of the cantilever and reflected into an array of photodiodes, similar to those used in digital cameras. The distance of the deflection is accurately determined. Since the tip is scanned across the sample multiple times, these microscopes are referred to as the "scanning" microscopes.

2.4.3 Electrical Method

The capacitance method is an electrical technique commonly used in scanning probe microscopy. It relies on the principle of parallel capacitors. The capacitance between two conducting elements is directly proportional to their surface area and the dielectric constant of the medium between them. Conversely, it is inversely proportional to the distance separating them. In the context of scanning probe microscopy, the capacitance between a smooth disk surface and the surface being measured is influenced by the roughness of the latter.

2.5 Surface Tension

Surface tension is a captivating property exhibited by the surface of a liquid, arising from the cohesive forces among the molecules.

2.5.1 Definition

Surface energy is the amount of energy possessed by a surface particle relative to an internal particle. The pulling force of the interaction between any two adjacent parts of a liquid surface is perpendicular to the unit length boundary. The formation of surface tension is related to the unique stress state of molecules in the thin layer of the liquid surface. The presence of surface tension gives rise to a range of the distinctive phenomena observed in daily life.

2.5.2 Measurement

Accurately measuring surface tension is a challenge due to the significant impact of even small quantities of surfactants or impurities with value. Inorganic liquids typically exhibit high surface tensions compared with organic liquids. To measure surface tension, a commonly employed method is the "wire frame" technique. This method involves suspending a rectangular wire frame into the liquid and applying an upward force, F_{up}, to counterbalance the downward force exerted by surface tension, T. The objective is to achieve equilibrium between the applied upward force and the force resulting from surface tension. This equilibrium is expressed by formula (2-1) or (2-2):

$$F_{down} = 2Tl = F_{up} \tag{2-1}$$

$$T = \frac{F_{up}}{2l} \tag{2-2}$$

When the measuring unit (Wilhelmy plate) comes into contact with the liquid surface, the liquid wets the Wilhelmy plate upwards. The surface tension acts along the perimeter of the plate and the liquid pulls in the plate.

2.5.3 Adhesion Theory

Adhesion theories encompass various principles, including adsorption and wetting, diffusion, donor/acceptor or electrostatic interactions, as well as the mechanical interlocking of the adhesive within the irregularities of the substrate. Recent research has made significant contributions to these theories. The adsorption theory highlights the crucial aspect that when the adhesive and substrate become in contact, the attractive force will come into play between them.

2.5.4 Adhesion Mechanism

The mechanical theory focuses on the interlocking mechanism between the adhesive and the rough surface of the substrate. This theory suggests that the enhanced adhesion occurs through the increased plastic energy dissipation during fracture within the bulk adhesive.

The electrostatic theory highlights electrical phenomena, such as sparking, observed during the destruction of an adhesive bond. It also considers the transfer of the electrostatic charge between the adhesive and substrate as a contributing factor to adhesion. The diffusion theory has garnered significant interest, especially with the development of the reptation theory of the polymer chain dynamics.

2.6 Surface Contact

The first analysis of the deformation and pressure at the contact of two elastic solids with geometries defined by the quadratic surfaces is due to Hertz (1882) and these contacts are referred to as Hertzian contact.

1. Sphere on a sphere

We address the issue of elastic deformation involving two spheres with radii R_1 and R_2 that are in solid contact with an applied normal load. The contact area is circular, with a radius R, and the contact pressure forms an ellipse with $P(r)$ at a radius r within the contact zone. The radius of the contact area is determined by:

$$r=\sqrt[3]{\frac{3\pi WR^*}{4E^*}} \tag{2-3}$$

$$E^*=\frac{1}{\frac{1-\nu_1^2}{E_1}+\frac{1-\nu_2^2}{E_2}} \tag{2-4}$$

$$R^*=\frac{1}{\frac{1}{R_1}+\frac{1}{R_2}} \tag{2-5}$$

Where W is the applied load, R_1 and R_2 are the radii of two contact spheres, v_1 and v_2 are the Poisson ratios of two contact surface materials, E_1 and E_2 are the elastic modulus of two contact materials, respectively. The maximum contact pressure ($P(r)_{max}$) at the center of the circular contact area is:

$$P(r)_{max}=\frac{3}{2}\frac{W}{\pi r^2} \tag{2-6}$$

2. Sphere on a rigid flat

Contact between a sphere and a rigid flat surface involves the interaction between a sphere and an elastic half-space. When sphere and flat contact, the force is linearly proportional to the indentation depth. The half-width r of the rectangular contact area is determined as:

$$r=1.109\sqrt[3]{\frac{WR}{E}} \tag{2-7}$$

Where W is the applied load, R is the radius of the sphere, and E is the elastic modulus of the sphere.

The maximum contact pressure along the center line of the rectangular contact area is:

$$P(r)_{max}=0.388\sqrt[3]{\frac{WE^2}{R^2}} \tag{2-8}$$

3. Cylinder on a cylinder

The contact surface of two cylinders under load W is rectangular when there is no elastic contact. Mark the width of the contact surface as $2a$, the length as b, and the contact pressure as P, then:

$$a=\sqrt{\frac{4WR^*}{\pi E^*}} \tag{2-9}$$

$$P=\frac{2W}{\pi ab}\sqrt{1-\frac{x^2}{a^2}} \tag{2-10}$$

Among them:

$$E^*=\frac{1}{\frac{1-v_1^2}{E_1}+\frac{1-v_2^2}{E_2}} \tag{2-11}$$

$$R^*=\frac{1}{\frac{1}{R_1}+\frac{1}{R_2}} \tag{2-12}$$

Where R_1 and R_2 are the radii of two contact cylinders, v_1 and v_2 are the Poisson ratios of the two contact surface materials, E_1 and E_2 are the elastic modulus of the two contact materials, respectively.

4. Cylinder on a rigid flat

Let's analyze the contact between two cylinders by envisioning a cylinder with the radius R_1 in contact with a rigid cylinder of infinite radius, and the following contact surface width $2a$ and contact pressure P are obtained by:

$$a=\sqrt{\frac{4WR_1(1-v_1^2)}{\pi E_1}} \tag{2-13}$$

Where R_1 is the radius of the cylinder, v_1 is the Poisson ratio of the material, and E_1 is the elastic modulus of the material.

2.6.1 Solid-Solid Contact

For two solid surfaces in contact, the interfacial bond may be stronger than the cohesive bond, particularly in a material with low cohesion. If two surfaces are placed together, due to surface roughness, the actual area of contact is usually significantly smaller than the geometric area. Adhesion is affected by the real area of contact, which varies on the basis of the normal load, surface roughness, and mechanical properties.

2.6.2 Solid-Liquid Contact

Generally, any liquid that wets or has a small contact angle on (hydrophilic) surfaces condenses from vapour on surfaces, either as a bulk liquid or in the form of an annular-shaped capillary condensate within the contact zone. The liquid film may also be deliberately applied for lubrication or other purposes. Adhesive bridges form around contacting and near-contacting asperities due to surface energy effects in the presence of a thin liquid film. The presence of these liquid films of the capillary condensates or the pre-existing films of the liquid can significantly increase the adhesion between solid bodies. When separating two surfaces, the viscosity of the liquid results in an additional attractive force, a rate-dependent viscous force, during the separation process.

2.6.3 Contact Angle

The contact angle appears to be directly proportional to the potential energy parameter between the liquid and solid interfaces. When a water droplet is absorbed by a monolayer water film, it exhibits a finite contact angle. This contact angle is determined by the surface energy between the monolayer water film and the bulk liquid water film. The variability in contact angle observed on different platinum crystal lattice structures can be explained by the characteristics of this monolayer water film.

The interaction among solid, liquid, and vapour phases, or more simply, the contact phenomena between a liquid and a solid surface, plays a crucial role in phase-change heat transfer. In recent years, the study of liquid-solid contact phenomena in the nanoscale systems has gained significant importance, particularly in nanotechnology applications such as the wetting of the catalyst metals in fuel cell electrodes.

2.7 Brief Summary

This chapter introduces engineering surface and interface, covering the physical and chemical properties of the surface, the definition and measurement of surface roughness, surface tension, and surface contact.

3 Friction

Friction is the tangential resistance force that arises at a sliding interface within a dynamic system. It is classified into two types: static friction and dynamic friction. Static friction occurs when there is no relative motion between two objects. Dynamic friction, also known as kinetic friction, occurs when two objects in contact are in relative motion and rub against each other.

3.1 Introduction

Friction is governed by three fundamental mechanisms: adhesion, ploughing, and deformation. These mechanisms collectively determine the resistance encountered when two surfaces come into contact and slide against each other. Understanding these mechanisms is crucial for comprehending the underlying principles of friction and its impact on various applications.

3.1.1 Definition of Friction

Friction is the force that opposes or resists the relative motion, or the tendency for relative motion, between two contacting surfaces. When one solid body slides over another, there is a resistance to the motion, which is known as friction. Friction plays a crucial role in various applications, such as belts, brakes, couplings, clutches, gears, bearings, and seals, where either increased or reduced friction is necessary. Solid friction includes static friction and kinetic friction. Fluid friction differs from solid friction and typically offers low resistance, primarily due to the motion of gas molecules or liquids. The friction between solids and fluids is known as fluid friction.

3.1.2 Origin of Friction

Amonton law states that the friction force F is proportional to the load N or the weight of the moving object, where the coefficient of friction is defined as the ratio of F to N:

$$\mu = F/N \tag{3-1}$$

The friction force is considered independent of the contact area. Coulomb proved the nominal contact area and sliding velocity are independent of the friction force. Bowden and Tabor considered that the friction force originates from the shear and plastic deformation of the adhesive nodes, which is proportional to the real contact area, and noted that the real contact area is rather small compared with the nominal contact area, but it tends to increase with the increase of normal load, leading to an increase with friction force.

3.1.3 Laws of Friction

The friction laws are formulated for the case of dry friction as follows:

(1) Amonton First Law: Friction force is directly proportional to the applied load between surfaces.

(2) Amonton Second Law: Friction force is independent of the apparent area of contact between solids.

(3) Coulomb's Law: Kinetic friction is independent of the sliding velocity.

3.2 Sliding Friction

Sliding friction refers to friction that occurs when the contact surfaces of two objects slide or have a tendency to slide against each other.

3.2.1 CoF

CoF is a value that quantifies the relationship between the friction force between two objects and the normal reaction force exerted between the objects involved. CoF is a dimensionless scalar that is defined as the ratio of the friction between two objects to the positive pressure that presses them together. In general, the coefficient of static friction is typically higher than that of dynamic friction.

3.2.2 Friction in the Lubricated Sliding

Lubricated sliding friction is a significant highly and intricate area of research. The sliding contact of elastic solids with smooth surfaces in viscous fluids with simple rheology is a well-explored problem. At the interface between a solid and a fluid, the fluid molecules "stick" or adhere to the solid, causing the fluid and solid velocities to coincide along the solid walls. The interaction potential between the solid atoms and the fluid molecules exhibits minimal lateral (atomic) corrugation, allowing slip to occur at the interface. Fluid dynamics is typically described by the Navier-Stokes equations, which serve as the foundation for understanding the behaviour of fluids.

When analyzing fluid flow between closely spaced solid walls, the Reynolds equations are often employed, derived from the Navier-Stokes equations with the assumption of fluid slip. Overall, the study of lubricated sliding friction delves into the intricate interplay between surface roughness, fluid rheology, and the dynamics of fluid-solid interactions.

3.2.3 Dry Friction

Dry friction is the force that opposes the motion of one solid surface sliding against another. It consistently acts against the relative motion between the surfaces and can either impede or initiate the movement of objects.

3.3 Rolling Friction

Rolling friction is a specific type of force that acts against the motion of a rolling object. It arises when an object rolls over a surface, creating resistance to its movement. This resistance is primarily caused by the deformation and interaction between the object and the surface, leading to energy loss and a deceleration of the rolling body.

3.3.1 Laws of Rolling Friction

When an object rolls on a surface, several phenomena occur: the object undergoes deformation at the point of contact with the surface, and the surface also experiences deformation at the point of contact with the object. This interaction generates motion beneath the surface. The magnitude of rolling friction depends on various factors, including the quality of the rolling object and the surface, the applied load, the diameter of the rolling object, and the surface area in contact. The coefficient of rolling friction represents the ratio of the rolling friction force to the total weight of the object.

Rolling friction is classified into three basic forms:

(1) Free rolling, known as pure rolling, where the rolling element moves in a straight line along the plane without any restrictions.

(2) Driven rolling, where the rolling elements are subjected to braking or driving torque, leading to normal pressure and tangential traction.

(3) Rolling in a groove, occurs when the geometric shape of the two rolling surfaces, and causes unequal tangential velocities at each point of contact.

There are three laws governing rolling friction:

(1) The force of rolling friction decreases with increasing smoothness of surfaces.

(2) Rolling friction is expressed as a product of the applied load and a constant raise to a fractional power.

(3) The rolling friction force is directly proportional to the applied load and inversely proportional to the radius of curvature.

3.3.2 Mechanism of Rolling Friction

Rolling resistance arises from two main factors: the sliding of one contacting surface along the other and irreversible deformation of materials in contact.

This sliding between the contacting surfaces contributes to rolling resistance. It leads to energy loss and hinders the smooth rolling motion of the object. Additionally, the irreversible deformation of the materials involved further contributes to the overall resistance experienced during rolling.

3.3.3 Measurement Methods

On a perfectly horizontal surface, observing the motion of a ball is a challenge unless the initial speed is very low or the observed path length is very long. This is because the effects of rolling friction are minimal at low speeds or over long distances.

In such cases, if the surface has a downhill slope, the ball accelerates, while on an uphill slope, it decelerates. These deviations make it difficult to accurately determine the coefficient of rolling friction.

3.4 Friction Theory

3.4.1 Bowden and Tabor's Simple Adhesion Theory

Adhesion between an elastic solid and a hard randomly rough substrate may take into account that partial contact may occur between the solids on all length scales. The normal load on the contact area is given by:

$$W = A_r \sigma_{xy} \tag{3-2}$$

Where A_r is the total and actual contact area, σ_{xy} is the compressive stress of the soft material in the friction pair, W is the applied outward load. This equation illustrates that the actual contact area A_r is proportional to the load W. The strong cold welding occurs at the contact, along with the tangential force F. Relative sliding causes shear at the nodes, resulting in resistance from the adhesive part $F_{\text{adhesion}} = A_r \tau_b$, resistance of the deformed part of F_E:

$$F = A_r \tau_b + F_E \tag{3-3}$$

Then:

$$F_a = A_r \tau_b \tag{3-4}$$

The expression of the coefficient of friction is given by:

$$\mu = F/W = \tau_b/\sigma_{xy} \tag{3-5}$$

Where τ_b is the shear strength (Pa) of soft material. Assuming the shear strength $\tau_b = \tau_c$, the critical shear strength is the shear stress when the material yields. The formula (3-5) can be written as:

$$\mu = \tau_c/\sigma \tag{3-6}$$

The stress state in the contact zone is complex.

For the easy quantification, equation (3-5) can be simplified as:

$$\mu = \tau_c/H \tag{3-7}$$

In this formula, the flow stress σ_y of the metal is substituted with the Vickers hardness H (Pa) of the soft material. And for most metals, τ_c is about 1/5 of the Brinell hardness of metals. So theoretically $\mu = 0.2$. The actual situation is $\mu > 0.5$ in the air.

Equation (3-5) ignores the influence of the increase of the real contact area under the action of tangential force and the influence of prestress caused by loading. For metals that are not work-hardened, the shear strength of the interface is approximately equal to the critical shear stress of the metal. This indicates a necessity to refine the adhesion. Bowden and Tabor's inelastic adhesion theory establishes a constant kinetic coefficient of friction.

According to the above-mentioned formula (3-4), the tangential force $F_a \propto A_r$, under the conditions of plastic contact, $A_r \propto W$ becomes $F_a \propto W$. This leads to a conclusion that the frictional force is proportional to the normal load and has nothing to do with the nominal contact area. Hence, it is affirmed the validity of the first and second laws of tribology (Amontons' law).

3.4.2 Modified Adhesion Theory

The disadvantage of the primary adhesion theory is that it does not consider the stress state of the bonding point, that is, the mutual relationship between the tangential force and the normal stress. When a micro convex body is in contact with a smooth surface, under the combined action of positive pressure and friction, the convex peak undergoes plastic deformation, and the flow of metal increases the contact area from A_r to $A_r + \Delta A_r$. The plastic deformation of the convex peak should satisfy the following condition:

$$\sigma^2 + \alpha\tau^2 = K^2 \tag{3-8}$$

Where K is the deformation resistance of the material, α is the coefficient, $\alpha > 1$. The larger the α, the greater the effect of the tangential force is. Various experiments have proved that as sliding progresses, the adhesion points grow in size.

In a frictionless unidirectional compression state, the friction stress $\tau = 0$, then:

$$K = \sigma_s \tag{3-9}$$

So,

$$\sigma^2 + \alpha\tau^2 = \sigma_s^2 \tag{3-10}$$

The adhesion theory for the existence of the contaminant film (or interface film) believes that the shear strength τ of the fouling film is used to replace the shear strength τ_m of the material itself. In general, $\tau_i < \tau_m$, so equation (3-10) can be expressed as:

$$\sigma^2 + \alpha\tau_i^2 = \sigma_s^2 \tag{3-11}$$

This equation also considers the combined effects of positive pressure and tangential force. When the tangential force of the contact pair is lower than τ_i, the bonding point continues to grow as it would on a clean surface, and when the tangential force exceeds the shear strength of the interface film, the growth of the bonding point stops, and the contaminant film is sheared. As a result, the macroscopic slippage occurs, leading to an increase in the real contact area as described by the second term in equation (3-10), which is significantly larger than the first term, then:

$$\alpha\tau^2 = \sigma_s^2 \tag{3-12}$$

3.4.3 Deformation Theory

Friction dissipates the mechanical energy through deformations during sliding. The slip-line field theory analyzes deformations in rigid, perfectly plastic materials. Plastic deformation is crucial for a viable friction theory, while elastic deformation alone is inadequate. The deformation theory addresses the asperity deformation as a unified problem with well-defined geometry, stress equations, and boundary conditions. Slip-line fields depict the planes of shear deformation, maintaining the material continuity.

3.4.4 Plow Theory

Ploughing occurs when two bodies in contact have different hardness. The asperities on the hard surface may penetrate into the soft surface and produce grooves on it, if there is relative motion. There are two basic reasons for ploughing: ploughing by surface asperities and ploughing by hard wear particles present in the contact zone.

The ploughing term of friction is an inelastic deformation term. When a hard and rough surface slides on a soft surface, the friction resistance is mainly caused by the "ploughing" of the asperities on the hard surface over the soft surface. CoF is expressed as:

$$\mu_p = F_T/W = A_T/A_r \tag{3-13}$$

In fact, the value of A_T/A_r depends on the shape of the conical indenter. According to the geometric relationship, it is obtained by:

$$\mu_p = \frac{2}{\pi}\cot\varphi \tag{3-14}$$

Where φ is the apex semi-angle of the conical asperity. The main points of the adhesion theory are as follows: there is a high pressure on the surface contact points, causing the contact points to be cold-welded or stuck together. The adhesive knots formed are subject to shear when the surfaces slide against each other. The formation and cutting of the adhesive points alternate on the surface, constituting the adhesion part of friction. If the materials of the friction pair have different hardnesses, or there are hard protrusions on the surface, the hard protrusions will form furrows on the surface of the soft material like a plough, forming a deformed part of the frictional force. In general, the adhesive part and the furrow part are not independent and unrelated. Regardless of the interaction of these two aspects, the total friction force is the sum of two parts, that is:

$$F = F_{\text{adhesion}} + F_{\text{deformation}} \tag{3-15}$$

$$\mu = \mu_{\text{adhesion}} + \mu_{\text{deformation}} \tag{3-16}$$

At is the horizontal projection of the asperity contact area. Ar is the vertical projection of the asperity contalt area.

3.5 Friction Model and System

Friction force models play a fundamental role in the simulation of mechanical systems.

3.5.1 Contact Mechanics and Multi-Asperity Model

The first study of contact mechanics was conducted by Hertz, who provided the solution for the frictionless normal contact of two elastic bodies of a quadratic profile. In 1957, Archard extended Hertz's solution to account for the contacts between rough surfaces, demonstrating that in a simple fractal-like model where small spherical bumps (or asperities) are distributed on the top of large spherical bumps, the area of real contact varies nearly linearly with load. A similar conclusion was reached by Greenwood and Williamson, who again assumed asperities with spherical summits (of identical radius) with a Gaussian distribution of heights. A general contact mechanics theory was developed by Bush.

The Greenwood and Williamson model describes the multi-asperity contacts of two real rough surfaces. When two real surfaces are separated by a distance d (defined from the mean of asperity heights), the number of contacting asperities (n) is determined by:

$$n = N\int_{\bar{d}}^{\infty} \varphi(\bar{z})\mathrm{d}\bar{z} \tag{3-17}$$

Where N is the total number of asperities, σ is the standard deviation of asperity peak heights, $\bar{z}=z/\sigma$ is the dimensionless height coordinate measured from the mean of asperity heights, $\varphi(\bar{z})$ is the probability density of asperity peaks, and $\bar{d}=d/\sigma$ is the nondimensional separation between the two surfaces. The general relationship between the normal load (P) and the deformation (u) of two spherical asperities in contact is given by the Hertz contact theory, expressed as:

$$P=\frac{4}{3}E^{*}R^{\frac{1}{2}}u^{\frac{3}{2}} \tag{3-18}$$

Where R is the composite radius of the curvature of the asperity tips, and E^{*} is the composite Young's modulus of two materials. Assuming one of the surfaces to be rigid and flat, the composite Young's modulus is given by $E^{*}=E/(1-v^{2})$, and the effective shear modulus becomes $G^{*}=2G$, where v, E, G are the Poisson's ratio, Young's modulus, and shear modulus of elastic body.

3.5.2 Finite and Boundary Element Methods

Finite and boundary element methods are well-established numerical techniques to solve various types of boundary value problems (BVPs) encountered in real-world systems. The finite element method mainly deals with the approximation of a Hilbert space as a Sobolev space with a finite-dimensional subspace. It also encompasses error estimation solution on the function space with a solution on the finite-dimensional space. This approach is based on a procedure like the partition of domain (Ω) in which the problem is posed into a set of simple subdomains, known as elements, which are often geometric shapes such as triangles, quadrilaterals and tetrahedra.

3.5.3 Atomistic Methods

Molecular Dynamics (MD) was initially developed to explore the interactions among hard spheres but has expanded to investigate various physical, chemical, and mechanical phenomena. Classical MD predicts the atomic motion using Newtonian or Langevin equations of motion, employing the interaction potentials. Classical MD focuses on atomistic dynamics and relies on the interaction potentials, also known as Force Fields (FFs). Although these FFs have their limitations, they offer fundamental insights into phenomena like the breakdown of continuum contact mechanics at the nanoscale and the phase behavior of fluids in confinement.

3.5.4 Multiscale Modelling

Multiscale modelling involves the interaction of different models at different scales to enhance the understanding of the complex phenomenon. Spatially multiscale problems, such as contacts between rough surfaces, involve multiple scales of the geometrical features. Temporally multiscale problems involve the release of long-built stresses in a short time. Spatially multiscale problems are more complex than temporally multiscale ones. Multiscale contact problems can be solved through classical or multiscale models, where the information exchange occurs between different scales. The study of multiscale rough contact has inspired numerous theoretical and computational investigations.

3.6 Fluid Friction

Friction is a phenomenon not only observed in solids but also in fluids, encompassing both liquids and gases. This resistance to motion within or between fluids is termed fluid friction. This force occurs between layers of fluid and restricts the movement of the medium moving through the fluid or within itself.

3.6.1 Examples

Several examples of fluid friction are listed below:

(1) Movement of fish in the water: The streamlined body of a fish minimizes friction, adding its swimming ability in water.

(2)Flow of honey: Honey always flows slowly due to the internal friction among its molecules.

(3) Lubrication: The lubricants applied to hinges reduce friction, facilitating smoother movement.

3.6.2 Laws

There are three fundamental laws that govern fluid friction:

(1) First law: Increasing the surface area of contact between the fluid and the surface leads to a corresponding increase in fluid friction.

(2) Second law: Elevating the object's velocity per unit distance increases the frictional force acting on it.

(3) Third law: Fluids with high coefficients of fluid friction exhibit a great frictional force coefficient.

3.6.3 Viscous Friction

Viscous friction is a type of frictional force that occurs in fluids, such as liquids or gases. It is also referred to as fluid friction or internal friction. Viscous friction primarily depends on the viscosity of the fluid, which measures its resistance to flow. The magnitude of viscous friction depends on the velocity of the object and the surface area in contact with the fluid. Viscous friction is described by the equation:

$$F = \eta A v \tag{3-19}$$

Where F is the frictional force, η is the dynamic viscosity of the fluid, A is the surface area, and v is the velocity of the object.

3.7 Friction of Materials

Friction of materials is an important property in various mechanical applications.

3.7.1 Friction of Metals

Friction of metals is primarily influenced by surface cleanliness. Clean surfaces exhibit high adhesion, friction, and wear. Metallic bonds of the interface facilitate the transfer of metal layers, causing wear debris. Contamination or oxide layers decrease friction levels. Factors such as load, sliding velocity, and temperature influence the formation of transfer layers, surface friction heating, and mechanical properties of the materials.

3.7.2 Friction of Ceramics

Friction of ceramics differs from metals due to their bonding nature, characterized by covalent and ionic forces. Ceramics have limited plastic flow and low ductility compared with metals but exhibit good mechanical strength, oxidation resistance, and resistance to corrosive environments. The frictional behavior in ceramics is influenced by properties such as fracture toughness, sliding speed, applied load, and temperature. An increase in fracture toughness tends to reduce friction by dissipating energy at the contact interface. However, the increased normal load leads to high friction and wear due to crack formation. Temperature and velocity affect ceramics through the formation of tribochemical films and surface fracture. Friction increases with temperature as the material softens, resulting in a small contact area and great friction.

3.7.3 Friction of Polymers

The friction behavior of polymers is different in comparison to the metals and ceramics. Polymers typically exhibit very low CoF due to their low elastic modulus and strength. They lack rigidity and strength and tend to flow under modest pressures and temperatures. Hence polymer composites are often employed to improve their friction behaviors for various applications. The primary forces influencing polymer friction behaviors include adhesion, deformation, and elastic hysteresis. Surface roughness and normal load affect CoF in polymers. Additionally, factors such as asperity deformation, sliding velocity, and temperature impact the friction behavior of polymers.

3.8 Brief Summary

This chapter introduces the concept of friction, the law of friction, friction theory, the measurement of friction, and emphasize the importance of friction.

4 Wear

Wear is a significant aspect of tribology, which is the science of interacting surfaces. It refers to the surface damage that occurs when solid surfaces come into contact and undergo relative motion.

4.1 Introduction

Wear is defined as the damage of a solid surface, generally involving progressive loss of material due to the relative motion between that surface and one or more contact substances.

4.1.1 Significance of Wear Research

Wear processes can be categorized into low, mild, and severe states. Low wear involves localized deformation and ploughing, while mild wear exhibits micro-cracks and localized fractures. Severe wear occurs under high-stress levels, leading to crack propagation and macroscopic fractures. Wear-related failures account for a considerable proportion (60% to 80%) of mechanical part failures, alongside fractures and corrosion. Understanding wear mechanisms and implementing measures to improve wear resistance is crucial for material and capacity savings, enhancing equipment performance, prolonging service life, and reducing maintenance costs, benefiting the national economy.

4.1.2 Contents of Wear Research

The purpose of studying wear is to find out the patterns of changes and the factors influencing wear by investigating and analysing various wear phenomena. Thus, it is necessary to seek the identification of measures to control wear and improve wear resistance. Generally, the main contents of wear research include identifying wear types under different working conditions, examining the conditions, characteristics, and evolutionary patterns of specific wear types, analyzing factors affecting wear such as friction pair material, surface roughness, lubrication state, environmental factors, sliding speed, load, temperature, and other working conditions, developing wear models, proposing measures to improve wear resistance, and utilizing wear testing technology and analysis methods.

4.2 Types of Wear

There is a broad spectrum of terms available for describing wear mechanisms. Some common types of wear include adhesive wear, abrasive wear, surface fatigue, fretting wear, erosive wear, and corrosion wear, as illustrated in Fig. 4 - 1.

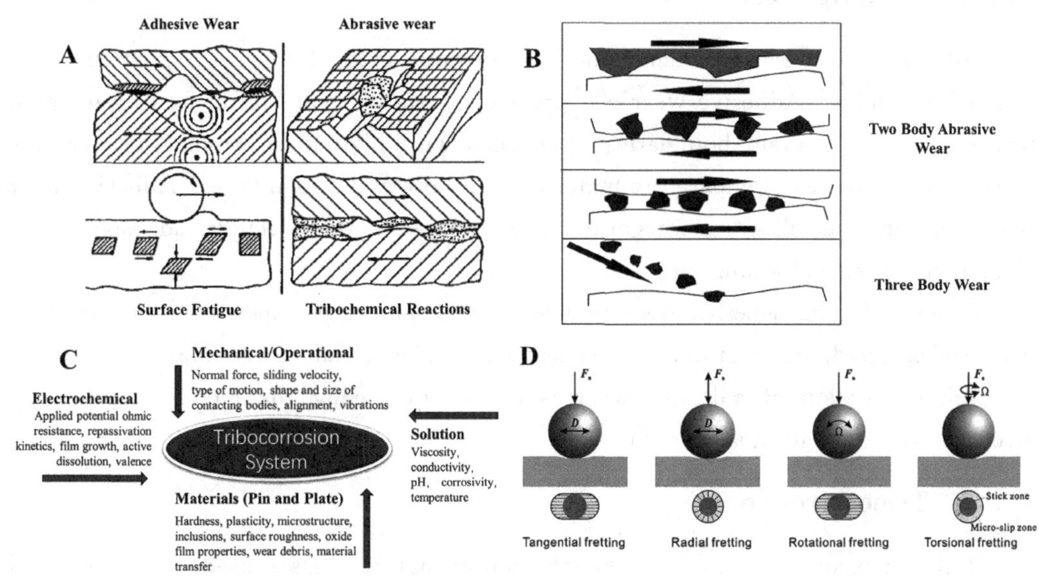

Fig. 4 - 1 Brief introduction of various types of wear

4.2.1 Abrasive Wear

In the process of friction, the material loss caused by the extrusion and movement of hard particles or hard micro-convex bodies on the surface of a solid friction pair is termed abrasive wear.

Abrasive wear is divided into three forms: two-body abrasive wear occurs when particles move along a solid surface, resulting in scratches or tiny furrow marks. Impact wear arises when particles move nearly perpendicular to the solid surface. The rough peak on the hard surface plays an abrasive role on the soft surface, which is also two-body wear, usually low-stress abrasive wear. Three-body wear occurs when external abrasive particles move between two friction surfaces, resembling a grinding effect, leading to plastic deformation or fatigue on the friction surface of ductile metal and brittle cracks or peeling on the surface of brittle metal. Factors affecting abrasive wear include hardness, strength, shape, sharpness, size of abrasive particles, and load.

Abrasive wear is governed by several mechanisms, such as micro-cutting, extrusion stripping, and fatigue failure. The Archard approach is used to estimate the volume of abrasive wear:

$$V = k\frac{W}{H}L \tag{4-1}$$

Where V is the volume removed, W, H, and L are the normal load, hardness of soft material, and sliding distance, respectively, and k is the wear coefficient typically ranges between 10^{-6} and 10^{-1}.

4.2.2 Adhesive Wear

Adhesive wear occurs when adhesive junctions created in the contact area are sheared by sliding. Adhesive wear is a prevalent form of wear, influenced by material properties such as grain boundaries. Materials with high grain boundaries generate elevated rates of wear compared to materials with low grain boundaries. Adhesive wear is divided into the following categories: slight adhesive wear, general adhesive wear, scratching wear, and gluing wear.

Factors affecting adhesive wear include material properties, material microstructure, load, sliding speed, temperature, environmental atmosphere, and surface film.

The calculation of adhesive wear is carried out using the model proposed by Archard, the same formual (4 - 1).

4.2.3 Tribocorrosion

Tribocorrosion refers to "the phenomenon of material loss caused by the chemical or electrochemical reaction between the surface material and the surrounding medium accompanied by mechanical action during the relative sliding of the dual surface of the friction pair". Tribocorrosion is often influenced by various material factors (material composition, structure, mechanical properties, and physicochemical properties), electrochemical factors (type, concentration, pH value of the corrosive medium), mechanical factors (load, speed), and environmental factors (temperature and pressure).

Based on the nature of the corrosive media involved, tribocorrosion can be divided into chemical tribocorrosion and electrochemical tribocorrosion. Research both domestically and internationally focuses on these factors: corrosive medium, mechanical factors, chemical composition, microstructure and properties of materials, and electrochemistry.

4.2.4 Fatigue Wear

When two contact surfaces undergo pure rolling or rolling sliding composite friction, small pieces of materials peel off in the local area of their interaction surface after multiple stress cycles under high contact compressive stress. This kind of surface fatigue is known as fatigue wear. Fatigue wear is divided into the following two types:

(1) Pitting. Pitting corrosion is characterized by the emergence of the initial cracks

appearing on the surface of the part, which gradually propagates, leading to fatigue failure.

(2) Spalling. In situations where the surface contact compressive stress is large and CoF is small, the initial cracks often initiate and expand beneath the surface. Most fatigue failures occur in this scenario, with material detaching in flakes and resulting in a large failure area. This form of fatigue wear is termed as spalling.

4.2.5 Impact Wear

Impact wear is produced either by erosion or percussion. Erosion occurs when the generated kinetic energy of solid particles in air or liquid streams interact with a surface, transferring kinetic energy. This kinetic energy generates contact stress, resulting in plastic deformation. Subsequently, erosion can cause cutting of the material or inter-section or the formation of cracks, thereby inducing wear of the material. This mechanism is a combination of adhesion, abrasion, surface fatigue, and fracture.

4.2.6 Fretting Wear

Fretting is a contact phenomenon that occurs between two mating bodies. Typical experimental conditions include factors such as frequency, medium, contact pressure (or normal force), and the relative displacement between the two contacting bodies.

In general, according to the relative motion directions, in the context of a ball-on-flat contact, fretting wear modes encompass the tangential, radial, torsional, and rotational modes.

4.3 Prediction of Wear

Wear is a common failure phenomenon in mechanical systems, causing the replacement of numerous components in many fields. The wear prediction model may include four components: ① interface contact modelling, ② mathematic coupling mechanism modelling, ③wear equation establishment, ④characterization of wear life.

4.3.1 Quantitative Indicator

Wear loss refers to the amount of material lost from the surfaces in relative motion during wear, and is divided into volume wear, mass wears, and wire wear.

(1) Wire wear amount Δh, volume wear amount ΔV, or mass wear amount ΔW. The wire wear amount Δh (μm) refers to the change in surface size during wear, which is measured in the normal direction of the wear surface. The wear quality ΔW is the amount of mass change before and after wear and is generally measured in mg. The wear volume is the amount of volume change before and after wear. The ΔV is small and usually expressed in mm^3.

(2) Wear rate. The ratio of the wear amount to the time or stroke at which wear occurs.

Wear rate can be expressed as: ①the amount of material wear per unit time, ②the amount of material wear per unit sliding distance, ③ the amount of material wear per revolution or reciprocating cycle.

(3) Specific wear rate. Specific wear rate is defined as the unit load (N) divided by the wear volume within one unit friction stroke (m) ($mm^3/N \cdot m$). The amount of wear is related to the sliding distance L, generally expressed as:

$$\frac{dV}{dt}=kpv \tag{4-2}$$

Where k is the wear coefficient of the material under certain working conditions, depending on factors such as lubrication, media conditions, the geometry and size of the contact surface, and the properties of the friction pair material. It is typically determined experimentally.

4.3.2 Wear Testing Methods

Various wear test methods are utilized to assess different types of wear. These tests evaluate qualities like thickness, porosity, adhesion, strength, hardness, ductility, chemical composition, stress, and wear resistance for quality control purposes. Non-destructive tests, such as visual inspection, penetrant dies, magnetic particle examination, and acoustic techniques, are commonly employed. Machine tools are used to conduct tests on both coated and uncoated cutting tools, subjecting them to wear parameters such as impact, shock, abrasion, adhesion, and hot corrosion.

4.3.3 Predictive Modelling

The advantage of simulation is that it has controllable variables. For a friction and wear phenomenon, the relevant parameters and constraints are deliberately selected to avoid the interference of other factors.

Simulation of abrasive wear. Surface roughness has a great influence on abrasive wear. The wear degree is related to the stress distribution. That is, areas with high contact stress experience more severe wear.

Simulation of adhesive wear. Holm model and Archard model are representative of adhesive wear simulation. The archard model is more representative of real-world adhesive wear, so most adhesive wear is modified on the basis of this model. These simulations establish adhesive wear models considering the effect on rough surfaces, where the wear criterion is whether surface micro convex bodies undergo the complete plastic deformation. Parameters such as contact area, wear rate, and wear coefficient are expressed as functions of factors including the normal load, surface roughness, elastic-plastic deformation parameters, surface energy, and material compatibility.

4.3.4 Wear debris analysis

Wear debris analysis is crucial for understanding the characteristics of a tribological system. Currently, research on the two-dimensional features of wear debris is well-developed, with the main hurdle in three-dimensional analysis lies in acquiring surface height data. Computer technology and mathematical methods are used to extract the characteristic parameters from wear debris for quantitative information and statistical recognition.

4.4 Wear of Materials

Materials undergo wear due to the mechanical action exerted on their surface, the environmental factors, and their physical and chemical nature.

4.4.1 Wear of Metals

Clean metallic surfaces are highly susceptible to friction and wear. Contamination on the material surfaces forms chemical films that reduce adhesion, thereby reducing the wear rate. CoF is influenced by three phenomena in the friction process: adhesion of smooth area of the sliding surface, ploughing caused by abrasive particles and hard roughness on the dual surface, and deformation of rough surfaces. Generally, the influence of ploughing and rough surface deformation on the total CoF is greater than that of adhesion.

Factors affecting dry sliding friction and wear behavior of materials are load, speed, and temperature field.

4.4.2 Wear of Ceramics

Ceramic materials possess high mechanical strength, corrosion resistance, and are prone to oxidation at elevated temperatures. These properties of ceramics result in low-area contacts which are helpful for reducing friction and wear at the surface interface. However, changes in mechanical deformation can increase fracture toughness at the interface, leading to an increase in wear rate at the interface.

Operating conditions such as applied load, sliding velocity, and environmental factors affect the wear rate of ceramics. Generally, the higher the hardness of the material, the better the wear resistance, however, the wear-resistance not only depends on hardness but also on factors like the bending-resistance strength or fracture toughness.

4.4.3 Wear of Polymers

Polymers offer several advantages over metals and ceramics, including stable chemical properties, strong corrosion resistance, low CoF and moderate wear rates, low relative density, high specific strength and remarkable vibration reduction effect. Wear mechanisms in polymers encompass adhesion, abrasion, and fatigue. The interface temperature of polymers and other solid lubricants is characterized as the function of normal pressure-sliding velocity (PV). Beyond the PV limit polymers can melt even at ambient temperature. High-temperature resistant polymers known for excellent performance mainly include polytetrafluoroethylene (PTFE), polyimide (PI), polyether ether ketone (PEEK), and polyphenylene sulfide (PPS).

4.4.4 Wear of Composites

Hybrid composites exhibit high wear resistance. The presence of large SiC particles allows them to bear a significant portion of the applied load, protecting the small alumina particles from wear. Soft second-phase particles in hard and robust matrix composites are also used in sliding applications, offering low CoF and adaptability in different operating conditions. The tribological properties of materials depend on the matrix, second-phase particles, as well as their size, shape, and concentration. Solid lubricant particles deform under sliding action, forming a soft interfacial film as they are extruded toward the surface.

4.5 Wear Control

Material wear processes encompass one or a combination of wear mechanisms, including abrasion, fretting, adhesion, fatigue, oxidation, and other tribo-chemical reactions.

4.5.1 Surface Modification

Surface modification technology utilizes chemical and physical methods to enhance the performance of machine parts or materials by altering the chemical composition or the microstructure of the surface. Mxenes, a new class of materials, have shown promise as reinforcements in polymers, metals, and ceramics.

Surface hardening is a technique used to enhance the surface hardness of metal materials, improving their resistance to wear and fatigue. Traditional methods like carburizing, nitriding, and boriding involve high-temperature element diffusion on the component surfaces. However, modern techniques such as plasma surface diffusion, laser surface alloying, and thermal spraying offer the efficient and environmentally friendly alternatives, capable of inducing fine microstructures or introducing additional elements to the surface for the improved performance.

Laser surface treatment uses high-power density laser beams to treat the metal surface in a non-contact way, and form a treatment layer with a certain thickness on the material surface, to change the structure of the material surface and obtain ideal properties. Laser surface strengthening technology is more suitable for iron and steel materials and cast iron materials.

4.5.2 Surface Coatings

Applying a wear-resistant material to the surface or creating a composite by embedding particles into the surface can be considered a good solution to the problem, which includes unique structures, the core-shell structure, amorphous nanocrystalline, and gradient multilayer coatings. A critical parameter of hardness/elastic modulus, H/E, is demonstrated to be a suitable factor for evaluating coating wear behaviors.

4.5.3 Bionic Coatings

Hierarchical surface engineering is initially inspired by natural phenomena and used to adjust the hydrophobicity and adhesion on surfaces of components. Lotus leaves give the famous hierarchical surfaces that exhibit two-tiered roughness resulting from the superposition of two roughness patterns at different length scales. So far, various techniques have been developed to create multi-scale surface roughness, such as nano-scratching, laser surface texturing, chemical etching, and end milling. The friction reduction caused by hierarchical surfaces is observed for materials such as titanium, sapphires, polymers, and composite materials.

4.5.4 Surface Texture

Surface texturing involves the controlled modification of topography to produce functional surfaces. These surfaces act as the efficient lubricant reservoirs, delivering lubricants and trapping wear debris at contact surfaces. Several mechanisms of surface texturing have been identified for tribological behavior, including ①the enhancement of load-carrying capacity by increasing hydrodynamic pressure over the surface texture, ②application of additional lubricants over the real contact area by the inlet-suction effect, ③reduction of the real contact area, ④storage of lubricants via the reservoir effect, ⑤capture of wear particles by the debris trapping effect. Combining dimples with a solid lubricant contributes to obtaining excellent tribological properties.

4.5.5 Properties of Wear-Resistant Materials

Key factors affecting material wear behaviors include testing conditions and intrinsic material properties. Materials with high hardness do not always show high wear resistance in practical applications. Both low and high structural stiffness scenarios displayed conflicting effects on the wear stability, whereas medium stiffness gave rise to relatively stable wear. The yield strength of steel was related to the susceptibility of specific wear mechanisms, including micro-cutting and micro-ploughing, with high wear coefficients. The ratio of hardness H to elastic modulus E, commonly known as the "plasticity index", is widely acceptable in determining the limit of elastic surface contact.

4.5.6 Wear Due to Chemical Reactions

The reactivity of the material surfaces and their interaction with the outer environment may lead to chemical reactions that can generate undesirable phenomena such as oxidation and corrosion. When these surfaces are subject to friction from rubbing, abrasion, and cracks, resulting in wear. There are also secondary types of wear like fretting and electrical induced wear.

4.5.7 Wear of Lubricated Contacts

The lubrication of contacting surfaces is determined by the Lambda ratio, which compares the film thickness to the roughness of the solid surfaces. A Lambda ratio above 3 indicates negligible contact and prevents mechanical wear. Ratios between 1 and 3 indicate a shared load between the lubricant and asperities, leading to adhesive wear. A Lambda ratio below 1 signifies serious contact, causing fatigue and adhesive wear. Lubricants can prevent chemical wear with a suitable composition and minimize abrasive wear by keeping surfaces clean. Fatigue and adhesive wear are the predominant types of wear observed in lubrication.

4.6 Wear Under Specific Conditions

Materials designed for extreme environments play a vital role in safeguarding individuals, structures, and the environment.

4.6.1 Corrugation

Rail corrugation has been a significant issue faced by the global railway industry in recent years. It refers to periodic irregularities that manifest on the rail's running surface, characterized by both long and short wavelengths. The formation and progression of rail corrugations bear resemblance to those observed on unsealed roads. The industry is concerned about this form of wear-induced damage, as there is currently no reliable alternative to regrinding surfaces. Friction modifiers are considered the most promising solution for addressing rail corrugation.

4.6.2 Electrical Field

Electric wear refers to the wear of friction pairs under the influence of an electric field during current flow. This research primarily focuses on high-speed railway systems, power transmission systems in urban transportation networks, industrial generators, carbon brushes, electrodes, and launch vehicle rectifier devices. These friction pairs undergo wear result of both high speeds and current flow.

4.6.3 Magnetic Field

Using a magnetic field to control friction and reduce wear is emerged as a promising approach in the tribological research. The increased temperature induced by the magnetic field is the main cause of the intensified surface oxidation. The application of a magnetic field facilitates surface oxidation and enhances dislocation mobility, consequently reducing wear rates. The mechanisms of oxidation and dislocation motion offer insights into wear behavior.

4.6.4 Wear Monitor

With the increasing scale and automation of production equipment, safety and reliability have become paramount concerns. Oil abrasive particle analysis technology is a non-disassembly monitoring and diagnostic method for modern equipment. The online wear monitor utilizes optics, electricity, and magnetic field theory to measure physical and chemical parameters of oil, such as dielectric constant and abrasive particle concentration, and assess wear conditions. Photoelectric monitors offer advantages like a simple structure, high precision, fast response, and non-contact operation. Magnetic plugs and iron removers use magnetic fields to attract ferromagnetic abrasive particles in oil, enabling wear state assessment based on the monitored particle deposition. Filter method monitors employ the magnetized steel fibre grids to detect changes in electrical signals caused by abrasive particles, allowing monitoring of both ferromagnetic and non-magnetic metal particles like copper, aluminum, and lead.

4.6.5 Wear-Rresistant Materials

Incorporating innovative materials and advanced processing techniques improves the strength, hardness, toughness, and anti-wear performance of surfaces, which is crucial for the durability, stability, and accuracy of mechanical parts at various scales. Aerospace components commonly are used by Al, Mg, Ti, and Ni-based alloys, with fretting wear being a major concern in applications such as bearing shafts, bolted connections, and blade-disc assemblies. Steel and aluminum alloys are widely used in transportation due to their mechanical properties and cost-effectiveness. Wind turbine blades, exposed to harsh conditions, require evaluation of fretting, abrasion, and erosion wear. Wear-resistant coatings like Co, Ni, and Ni_3 Al-based alloys offer protection against severe wear.

4.6.6 Wear State Detection and Identification

Wear state detection and identification effectively capture the performance changes of running machines. Oil analysis is widely used for wear and lubricant monitoring, providing direct insights into wear mechanisms. Wear states are determined by microscale wear mechanisms and macro-scale wear quantity. Wear state varies significantly among machines with different tribo-components and working conditions, necessitating specific prediction models for each machine. Wear is influenced by the material properties, lubricants, and conditions, leading to random wear properties in the short term, while exhibiting regular wear properties in the long term as a result of structural damage and material loss. Modelling wear requires the integration of both dynamic and statistical methods.

4.7 Brief Summary

This chapter outlines various types of wear, the wear of materials, and methods for wear calculations.

5 Lubrication Theory

Lubrication is a process or technique where a substance, known as a lubricant, is applied to reduce wear between one or two relatively moving surfaces under load. Lubricants can take various forms, including solids (MoS_2), solid/liquid dispersions (grease), liquids (such as oil or water), liquid-liquid dispersions, or gases.

5.1 Introduction

Reducing friction, wear, and the occurrence of a seizure by providing a suitable substance, such as oil between two surfaces in relative motion is called lubrication, and the substance, used for this purpose is called lubricant. Some solids are also used as lubricants, mostly in the form of coatings. It is very important in light of the finiteness of natural resources on the Earth. the effect of lubrication is usually more remarkable in reducing wear than in reducing friction.

5.2 Stribeck Curve and Lubrication Status

Stribeck curve is a fundamental concept widely known in the fields of tribology and lubrication. It illustrates the friction behavior in the lubricated contacts as a function of the lubricant viscosity, entrainment speed, and surface roughness (sometimes, just roughness).

5.2.1 Stribeck Curve and Status of Lubrication

It is known that the CoF of a journal bearing changes with operating conditions, as shown in Fig. 5 - 1. The vertical axis indicates the CoF (μ) and the horizontal axis denotes the bearing number $\eta U/P$, where F is the frictional force, P is the journal load, η is the coefficient of viscosity, and U is the circumferential velocity of the journal (the part of a shaft supported by a bearing). CoF (μ) has a minimum point as shown in the figure, where the value is very low, usually of the order of 0.001. For high values of the bearing number $\eta U/P$, the CoF μ increases along a straight line passing through the origin. The rate of increase is small. Conversely, with the decrease in the bearing number from the point of minimum CoF, CoF increases rapidly, but does not exceed a certain fixed value. The diagram is based on the meticulous experiments conducted by Richard Stribeck (1861—1950) in 1902, which lays the foundation for understanding the friction and lubrication behaviors in engineering applications.

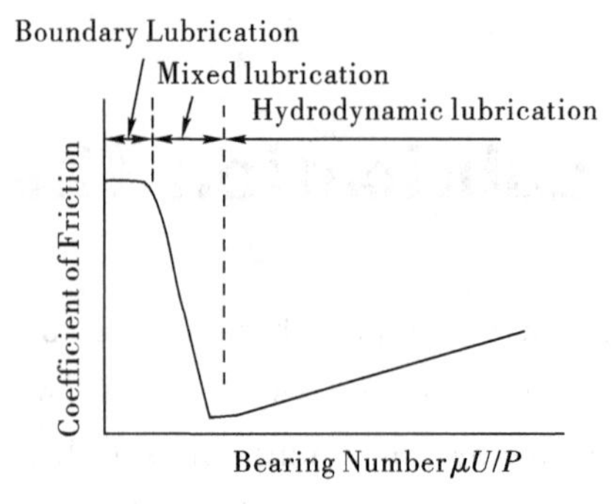

Fig. 5 - 1 Stribeck curve

The Stribeck Curve is a fundamental concept in tribology, delineating different lubrication regimes based on operating conditions. It is known as the boundary lubrication regime, in which solid/solid direct contact takes place causing deformation and breaking of asperities. The second regime is termed hydrodynamic regime, where the fluid film separation occurs between two surfaces. The third regime is called mixed lubrication regime, representing a mix of both boundary and hydrodynamic lubrication, where partial contact persists.

5.2.2 Boundary Lubrication

Friction and wear in boundary lubrication are determined predominantly by interactions between solids and between solids and liquids. Boundary lubricating films are created from surface-active substances and their chemical reaction products. Adsorption, chemisorption, and tribochemical reactions also contribute significantly to this process. One of the primary objectives of the lubricant development is the creation of the effective boundary friction layers under various geometric, dynamic, and thermal conditions. Boundary lubricating layers are created from surface-active substances and their chemical reaction products.

5.2.3 Fluid Film Lubrication

The surfaces are fully separated by a fluid lubricant film (full-film lubrication). This film is either formed hydrostatically or more commonly, hydrodynamically. From a lubricant's point of view, it is known as hydrodynamic or hydrostatic lubrication. Liquid or fluid friction is caused by friction resistance, because of the rheological properties of fluids. If both surfaces are separated by a gas film, Then, it is termed as gas lubrication.

Hydrodynamic Lubrication (HL) and Elasto-Hydrodynamic Lubrication (EHL) are lubrication regimes where a thin lubricant film is formed between two surfaces in relative motion. EHL usually occurs in non-conformal contacts and many machine elements, including rolling bearings and gears, relying on EHL for their operation. The existence of a sufficient fluid film to separate two surfaces under hydrodynamic conditions, as observed in a journal bearing, has been known since the work of Tower in 1883. The formation of such a film is facilitated by high pressure, which has two beneficial effects: firstly, it increases the lubricant viscosity in the contact inlet and, secondly, it elastically deforms and flattens the contacting surfaces, giving rise to the term elasto-hydrodynamic lubrication.

5.2.4 Mixed Lubrication

Mixed lubrication occurs when boundary friction combines with fluid friction. From a lubricant's technology standpoint, this form of friction requires sufficient load-bearing boundary layers. Machine elements which are normally hydrodynamically lubricated experience the mixed friction during startup and shutdown.

For roller bearings, one of the most important machine elements, it has been shown that the reference viscosity either of lubricating oils or of the base oils of greases is not sufficient to ensure the formation of protecting lubricant layers and the required minimum lifetime. Under the mixed friction conditions, it is important to choose the appropriate lubricant, one that enables the formation of tribolayers by the inclusions of anti-wear and extreme pressure additives.

5.2.5 Elastohydrodynamic Lubrication

Hydrodynamic calculation on lubricant films has been extended to include the elastic deformation of contact surfaces and the influence of pressure on viscosity. This enables the application of these elastohydrodynamic calculations to contact geometries beyond plain bearings.

In hydrodynamic bearings, the generation of fluid dynamic pressure mainly depends on the relative movement between the shaft neck and bush, the convergence gap, and the viscosity of the oil, ensuring support for the external loads. Conversely, hydrostatic bearings mainly rely on external oil pressure to support the load, with oil pressure increasing as the load improves. In most cases, the load improvement can lead to a decrease in the oil film thickness.

5.3 Boundary Lubrication

5.3.1 Definition

In 1922, British scholar Hardy first proposed the concept of boundary lubrication. Alongside Dubday, they noticed that when solid surfaces are in close proximity, the friction and wear characteristics of the surface are mainly adsorbed on the solid interface. As the bodies come into close contact with their asperities, the heat generated by local pressures cause a condition which is called stick-slip, and some asperities break off. Under elevated temperature and pressure conditions, chemically reactive constituents of the lubricant interact with the contact surface, forming a highly resistant tenacious layer or film (boundary film) on the moving solid surfaces. The film is capable of supporting the load, preventing significant wear or breakdown.

5.3.2 Significance

In the state of boundary lubrication, various factors come into play, including metallurgy, surface durability, physical and chemical processes such as physical adsorption, chemical adsorption, chemical reaction, corrosion, catalysis, temperature effects, and reaction time. The most important factor among them is the formation of a surface film on the metal in reducing the damage caused by solid-to-solid contact. Environmental media such as oxygen, water and substances that counteract surface activity can affect the formation of films. The effectiveness of boundary lubrication is determined by the physical properties of the films, including thickness, hardness, shear strength, cohesion, adhesion, melting point, and solubility of the film in the base oil.

5.3.3 Experimental Techniques for Characterization

The parameters in the boundary lubrication model need to be calibrated based on the experimental results. Successively, the tribofilm thickness, CoF, and wear depth predicted by the model are compared with those measured in the reciprocating experiments under various lubricant temperatures and applied loads. The wear depth of the liner segment is assessed using the surface profiler, with measurements taken from multiple locations across the wear track to determine standard deviation. The final wear depth is the average of these measurements.

5.3.4 Formation of Adsorption Film

The state of boundary lubrication is determined by the properties of the friction surface and the nature of the boundary. A boundary film is left on the upper metal surface, effectively reducing friction and wear. Boundary membranes are divided into two categories: adsorption membranes and reaction membranes, according to their formation process.

The boundary limit is formed by the polar molecules of the lubricant adsorbed on the friction surface, which is called the adsorption film. Adsorption membranes are further divided into physical adsorption membranes and chemical adsorption membranes. Physical adsorption occurs when substanccs are attracted to the solid surface through molecular forces, attraction to adsorb substances on the solid surface resulting in the formation of one to several molecular layers of molecules arranged in an oriented manner. Chemical adsorption film is the exchange of the valence electrons of the polar molecules of the lubricant with the electrons on the metal surface a generate chemical binding force, making polar molecules aligned and adsorbed on the surface, forming a surface film. The connection between the chemical adsorption film and the substrate is stronger than that of the physical adsorption film, leading to better lubricity.

5.3.5 Adhesion Between Lubricated Surfaces

Under the influence of the normal load, the contact of the relatively moving surface asperities increases, and the boundary film at some of the contact points ruptures, resulting in metal-metal contact and adhesion, and a small part of the surface is affected by the hydrodynamic pressure.

The frictional force is regarded as the shear resistance of the adhesion part of the shear surface and the shear resistance between the boundary film molecules and the liquid produced in the cavity between the micro-protrusions. In boundary lubrication, when the boundary film effectively lubricates, CoF is influenced by the shear strength inside the boundary film. Since it has much lower shear strength than metal during dry friction, CoF is also much smaller. In the state of boundary lubrication, the friction characteristics of the surface are improved by the presence of the boundary lubricant.

5.3.6 Adsorption Film

The adsorption film of boundary lubrication is a thin film of one to several molecular layers in which the polar molecules in the lubricating oil are aligned on the metal surface, akin to a fence with molecules arranged on it. When the concentration of polar molecules in the lubricating oil is sufficient to saturate the monomolecular layer adsorbed on the metal surface, the polar molecules are tightly arranged, and the cohesion among the molecules is very strong as if the molecules are gathered into a whole film with a certain bearing capacity. This arrangement effectively prevents direct contact between the two friction surfaces. When the friction pair slides, the surface adsorption film acts like two brushes sliding against each other, reducing friction and providing lubrication.

The Van der Waals force of the molecules on the solid surface influences the molecular layer of the oil film.

5.3.7 Boundary Lubrication at the Nanoscale

Nanoscale boundary lubrication is the study of lubrication at the nanometer scale, which occurs when two surfaces are near by and subjected to high loads and relative motion. Instead of, relying on fluid film formation or elastohydrodynamic lubrication, nanoscale boundary lubrication relies on the interaction of the surfaces at the atomic level, including the formation of chemical bonds and the rearrangement of atoms and molecules. This type of lubrication is observed when surfaces experience high loads and relative motion. At the nanoscale, the traditional lubrication mechanisms of fluid film formation and elastohydrodynamic lubrication are ineffective. Instead, nanoscale boundary lubrication relies on the interaction of the surfaces at the atomic level, including the formation of chemical bonds and the rearrangement of atoms and molecules. The effectiveness of nanoscale boundary lubrication depends on the properties of the surfaces, including roughness, chemistry, and crystalline structure.

5.3.8 Mixed Lubrication

This regime is between the full film elastohydrodynamic and boundary lubrication regimes. The generated lubricant film is not enough to completely separate the bodies, but the hydrodynamic effects are significant. Besides supporting the load, the lubricant may have to fulfil other functions as well, such as cooling the contact areas and removing worn products. While carrying out these functions, the lubricant is constantly replaced from the contact areas either by relative movement or by externally induced forces.

Lubrication is required for the correct operation of mechanical systems such as pistons, pumps, cams, bearings, turbines, and cutting tools. Without lubrication, the pressure between the surfaces in close proximity may generate excessive heat, leading to rapid surface damage, which in a coarsened condition may weld the surfaces together, causing seizure.

5.3.9 Wear

Under oil lubrication, wear is attributed to the failure and cracking of the boundary film. The oil film defect rate indicates the extent to which the oil film covers the surface, reflecting the probability of wear occurring.

The Kingsbury oil film defect rate β is related to the time at which the microprotrusions pass through the contact zone. The oil film defect rate β is defined as the ratio of the number of positions of the friction surface lubricant molecules to the total number of positions of the friction surface.

The boundary friction coefficient can be written as:

$$\mu_{BL} = \mu_{BF} + \beta(\mu_M - \mu_{BF}) \tag{5-1}$$

Where μ_{BL} is CoF of boundary lubrication; μ_{BF} is CoF of boundary film; μ_M is CoF of metal.

5.4 Fluid Lubrication

Hydrodynamic lubrication occurs when the motion of contacting surfaces, and the exact design of the bearing is used to the pump lubricant around the bearing, ensuing the maintenance of the lubricating film. The basis of the hydrodynamic theory of lubrication is the Reynolds equation.

5.4.1 Regimes of Fluid Film Lubrication

The governing principles include the conservation laws, conservation of mass, conservation of linear momentum, and conservation of energy. These are based on the classical mechanics and undergo modifications in quantum mechanics and general relativity. They are expressed through the Reynolds transport theorem. In addition to the above, fluids are assumed to obey the continuum assumption. Fluids are composed of molecules that collide with one another and with solid objects.

In 1886, Newton summed up the famous Newton's internal friction law through experiments and launched a new chapter in fluid viscosity research.

The equation is as the following:

$$\tau = \pm \eta \frac{\mathrm{d}u}{\mathrm{d}y} \tag{5-2}$$

η——friction shear stress, with its unit being Pa.

μ——dynamic viscosity, with its unit being Pa·s.

Reynold studied the fluid dynamic problems based on viscous the fluid mechanics equation and the flow continuity equation. The Reynolds equation is based on the Navier-Stokes equation and the continuity equation.

Then we can simplify the Navier-Stokes equation to the Reynolds equation with the above assumptions. The general formula of the Reynolds equation is:

$$\frac{\partial}{\partial x}\left(\frac{\rho h^3}{\eta}\frac{\partial p}{\partial x}\right)+\frac{\partial}{\partial y}\left(\frac{\rho h^3}{\eta}\frac{\partial p}{\partial y}\right)=6\left[\frac{\partial}{\partial x}(U\rho h)+\frac{\partial}{\partial y}(V\rho h)+2\rho(w_h-w_0)\right] \tag{5-3}$$

ρ——density;

p——pressurs;

U——velocity of upper surface;

V——velocity of lower surface.

The Reynolds equation is a two-dimensional second-order nonlinear partial differential equation.

5.4.2 Hydrodynamic Lubrication

Hydrodynamic lubrication is a state of lubrication that relies on the shape of the two sliding surfaces of the moving pair to form a layer of fluid film with sufficient pressure when they move relative to each other, thereby separating the two surfaces.

There exists a point in the fluid wedge, where the pressure gradient is zero, and the thickness of the fluid film at this point is h, then the following relationship can be obtained by sorting out:

$$\frac{\mathrm{d}p}{\mathrm{d}x}=\frac{U\eta}{2k}\left(\frac{h-\bar{h}}{h^3}\right)=6U\eta\left(\frac{h-\bar{h}}{h^3}\right) \tag{5-4}$$

Integrating this formula, we can obtain:

$$p=6U\eta\frac{h-\bar{h}}{h^{3}}\mathrm{d}x+c \tag{5-5}$$

$\bar{h}$——the film thickness of the maximum pressure point.

Usually, using the starting point and ending point of the fluid film pressure distribution curve, c and $\bar{h}$ can be judged, and the equation can be obtained.

There are two types of hydrodynamic lubrication. One involves the two solid surfaces with standing the pressure of the lubricating film, the gap between the two surfaces is a wedge, and the lubricant flows from the large mouth of the wedge to the small opening of the wedge, which is called a wedge gap effect. The other type occurs when the two solid surfaces are close to each other along the normal direction, causing the gap of the lubricant to be squeezed resulting in increasing pressure, which is called extrusion. However, in many practical situations, the two forms of hydrodynamic lubrication coexist.

5.4.3 Hydrostatic Lubrication

Hydrostatic lubrication refers to the use of an external fluid pressure source, such as an oil supply device, to deliver a certain pressure of fluid lubricant into the supporting oil cavity to form a fluid lubricating film with sufficient static pressure and bear the load, creating a lubrication state that separates the surfaces. This method is also known as external pressure lubrication.

5.4.4 Elastohydrodynamic Lubrication

The difference between elastohydrodynamic lubrication and hydrodynamic lubrication is that the consideration of the stiffness of the workpiece lubrication surface. Fluid dynamic pressure lubrication theory is feasible in the case of low pairs such as sliding bearings. Therefore, if the elastic deformation of the lubrication surface is not considered, a large difference is generated in the actual situation. This is the main difference between hydrodynamic lubrication and elastohydrodynamic lubrication.

Grubin's work is the earliest and most comprehensive understanding of the elastohydrodynamics theory. He gave an approximate solution to the problem of isothermal full-film elastohydrodynamics in heavy-line contact. In the calculation of the elastic flow, the dimensionless parameter is usually used to represent the oil film thickness, so the Grubin formula is:

$$H_0=1.95\frac{(G\bar{U})^{9/11}}{(\bar{W})^{1/11}} \tag{5-6}$$

H_0——$\frac{h_0}{R}$.

G——aE'.

$\bar{U}$—— $\frac{u\eta_0}{E'R}$.

$\bar{W}$—— $\frac{w}{E'R}$.

The oil film thickness calculated by the above formula is very close to the value of the test, so the Grubin theory is an important contribution to the development of elastohydrodynamics.

5.5 Lubricants

Lubricants are substances that reduce friction and wear between two surfaces, typically applied at the interface of the two surfaces.

5.5.1 Liquid Lubricants

The properties of lubricating oil are usually determined by certain methods, including appearance, density, viscosity, viscosity index, flash point, freezing point, pour point, and moisture. Liquid lubricants are used extensively in applications characterized by high speed and large loads. Liquid lubricants are comprised of base oil and some additives. Base oils include animal and vegetable oils, mineral-based lubricants, synthetic oil and water-oil emulsions. The functions of lubricants encompass keeping the moving parts separate, reducing friction, transferring heat, protecting against wear, preventing corrosion, and providing gas sealing.

5.5.2 Ionic Liquid(IL) Lubricants

Ionic Liquids (ILs) are salts of organic cations and inorganic or organic anions with a melting point below 100 °C. ILs have attracted considerable interest in different areas due to their unique physical properties, such as non-flammability, non-volatility, low melting point, outstanding thermo-oxidative stability, and dipolar structure. Besides, they also contain active elements, such as N, F, B, and P, which may potentially react with freshly generated metals during friction, forming antiwear compounds.

When a lubricant performs in a particular application, it should possess desirable viscosity, good lubricity, low volatility, non-flammability, highthermo-oxidative and chemical stability, and low toxicity. ILs exhibit a much lower CoF than rapeseed oil and mineral oil under the boundary lubrication and EHL regimes. The type, structure, and concentration of the ILs additives are different, resulting in different lubrication outcomes. Under boundary lubrication, high loads and very slow speeds produce extreme pressures, subjecting lubricant molecules to severe shear stress.

One mechanism involves the formation of a physical adsorption film between the ILs and the surface. The lubrication film generated by the reaction between the active elements in the ILs and the fresh surface are called the tribo-chemical reaction film. The ILs and the base oil molecules have synergistic or competitive adsorption on the surface during the formation of the lubrication film, thus, even if the content of the ILs as an additive is very small, it can achieve the same anti-friction results as that of the pure ILs.

5.5.3 Greases

Grease is the mixture of oil, thickener (soap), and additional lubricants (such as Teflon). Grease provides excellent protection against wear and tear and effectively seals against foreign particles. Grease can serve various purposes, including lubrication, corrosion protection, and sealing or filling applications. Grease maintains thicker films in clearances enlarged by wear and can extend the life of the worn parts that are previously oil-lubricated. Moreover, the thicker grease films also provide noise insulation.

5.5.4 Solid Lubricants

The solid lubricating properties of naturally occurring materials, such as graphite and molybdenite, are well known. Commonly used solid lubricants include graphite, polytetrafluoroethylene, and molybdenum disulfide. The lubrication effects of solid lubricants can be mainly divided into three types. Firstly, they can form a solid lubricating film on the friction surface, operating a mechanism similar to boundary lubrication. Secondly, soft metal solid lubricants leverage the low shear strength of soft metals for lubrication. Lastly, graphite serves as another type of solid lubricant.

5.5.5 Gas Lubricants

Gas also serves as a lubricant, where the friction surface between the moving components is separated by a gas film with sufficient pressure achieved through dynamic or static pressure. This film bears external loads, thereby reducing the friction resistance and surface wear during movement. Common gases used as lubricants include air, nitrogen, helium, carbon monoxide, and water vapour. The important feature of gas is its low viscosity, typically 1/100 to 1/1000 that of liquid, and it is compressible, and its density must be considered.

Gas bearings can be divided into gas dynamic pressure lubrication bearings, gas static pressure lubrication bearings, gas pressure film bearings, and gas foil bearings. The gas used as a lubricant is a viscous compressible fluid. The main properties include: transmission (diffusion, viscosity, and thermal conductivity), adsorption, and compressibility. These bearings also involve the presence of high friction during the startup and shutdown of the journal in loaded conditions. Researchers have endeavoured to enhance the load-carrying capacity and stability of foil bearings through various approaches over the years.

5.6 Brief Summary

Lubricants are commonly used for reducing friction and wear at interfaces. The Stribeck curve and lubrication state are introduced, followed by a brief overview of fluid lubrication, mixed lubrication, elastohydrodynamic lubrication, and boundary lubrication. Subsequently, the characteristics and principles of boundary lubrication, and the formation of boundary lubrication film are introduced in detail.

6 Superlubricity

Superlubricity is a state in which two sliding surfaces exhibit extremely low friction resistance to sliding, giving unusually super low friction. CoF is considered to be below 0.01 or approaching zero during sliding conditions.

6.1 Introduction

Friction between the mechanical parts causes surface wear and increases the energy consumption, which reduces the efficiency and the lifetime of mechanical systems. More and more researchers are searching for methods to reduce friction and decrease CoF, expecting to achieve a frictionless state in which CoF is almost zero, namely superlubricity. The concept of superlubricity was proposed by Hirano and Shinjo at the beginning of the 1990s to describe a theoretical sliding regime in which friction between two contact surfaces completely vanish. Theoretically, superlubricity is the realization of zero friction force. But in practice, due to the measurement precision limitations and other influencing factors, superlubricity is considered to occur when CoF is less than 0.01. Research on superlubricity is divided into two areas. One is in theory, where most works are focused on investigating the condition of superlubricity and the mechanism of superlubricity. The other one is in experiments where great efforts have been made for finding out more kinds of superlubricity materials.

6.1.1 Development

The first theoretical prediction of superlubricity in the crystalline interfaces was given for the infinite incommensurate contacts in 1983. Experimental confirmation of superlubricity dates back to 1993, with the observations on homogeneous MoS_2 interfaces. This is further supported by experiments on the nanoscale heterogeneous MoS_2/MoO_3 junctions, which exhibits the anisotropic friction characteristic of these systems. A decade later, the mechanisms of superlubricity in nanoscale graphitic contacts are undertaken, demonstrating controllable and reproducible superlubricity motion. Fig. 6 - 1 shows a brief introduce of superlubricity mechamism.

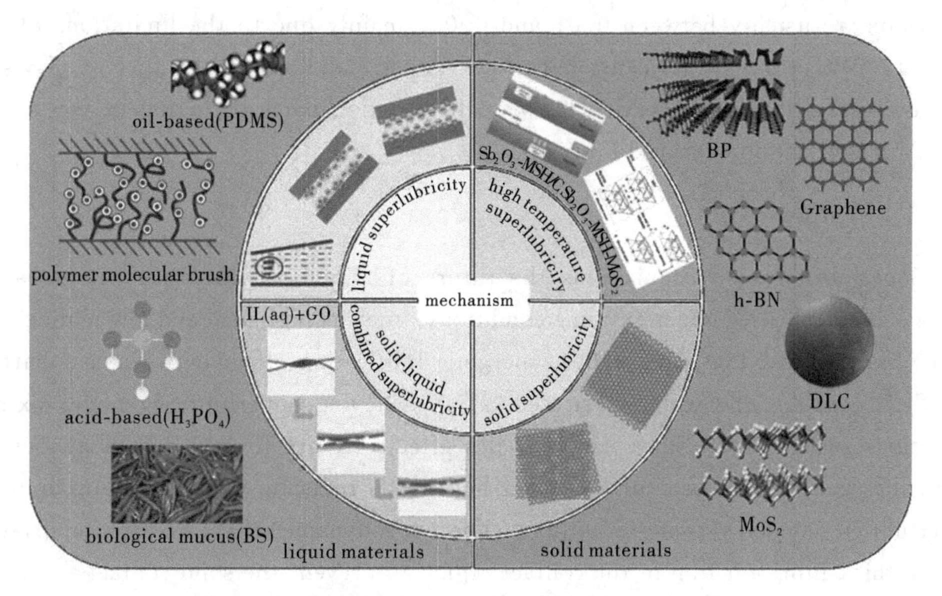

Fig. 6 - 1 A brief introduce of superlubricity mechamism

6.1.2 Challenge

Firstly, superlubricity achieved by the use of the heterogeneous contacts of rigid layered materials can be extended to the macroscale. Secondly, multi-contact configurations can serve as a venue for resolving some of the above mentioned problems, taking advantage of surface roughness and/or polycrystallinity to effectively transform macroscale junctions into a large collection of nanoscale contacts, even under high external loads and high sliding velocities. Thirdly, the way to reduce in-plane elasticity effects involves the deposition of two-dimensional material coatings on rigid surfaces and/or the usage of multi-layer stacks. Solid superlubricities such as diamond-like carbon (DLC), MoS_2, graphite, and CNx films exhibit the superlubricity phenomena under certain conditions when the friction surfaces directly contact each other. The other is liquid superlubricity, such as polymer brushes with water, ceramic materials with water, glycerol solution with acid or polyhydric alcohol, and some kinds of polysaccharide mucilage from plants.

6.2 Liquid Superlubricity

Liquid lubrication is a primary method for reducing friction. Unlike the mechanism described in solid superlubricity, the main source of friction in liquid lubrication is the internal friction of the fluid. Liquid lubricants mainly include two kinds: oil-based and water-based. Oil-based lubricants are characterized by a high viscosity and high coefficient of viscous pressure, so it is easy to form fluid lubrication between the surfaces of the friction pair. The lowest CoF corresponding to the traditional oil-based lubricants are usually between 0.01 and 0.05, mainly due to the limitations of their viscosity. Water-based lubricants have low viscosity and a small coefficient of viscous pressure, resulting in lubrication between the friction parts occurring in the form of boundary lubrication or mixed lubrication.

6.2.1 Water

Superlubricity is recognized as the future of tribology. However, it is hard to achieve superlubricity under extreme conditions, such as high load and low sliding speed at the macroscale. The remarkable synergetic lubricating effect between nanoparticles and Si_3N_4 enables the water-lubricated Si_3N_4 achieving superlubricity under extreme conditions successfully. Silica nanoparticles effectively are formed a homogenous film with silica gel on the worn surface under high load, reducing wear and maintaining the superlubricity under extreme conditions. The hydrodynamic effect is very important to form a thick lubricant film in the contact region to prevent the solid contact.

Water molecules adsorbed on polymers are formed hydration shells around them, leading to a hydration layer formed on the copolymer hydrogel surface. The superlubricity state is achieved in water lubrication and primarily related to the hydration effect caused by polymer chains. Because superlubricity is sensitive to the load and sliding speed, it is hard to achieve a super-low CoF under high load and low speed. Hence, additives such as ionic liquids, polyols, amine derivatives, acids, carbon nanomaterial, nanoparticles, and nanocomposites are added into water to improve the tribological properties. Superlubricity under high load and low speed is realized via the addition of silica nanoparticles, and the synergetic effect of the filling mechanism, film-forming mechanism, and double electric layer contributed to the excellent lubricating performance of amino-modified silica nanoparticles. Water-containing fluids have the potential to significantly reduce fluid friction further. The highly-loaded EHL contacts with a water-soluble glycerol as base oil with water content, ultra-low coefficients of friction between 0.005 to 0.010 are measured. The water-containing fluids show the significant potential for the gear applications in terms of the efficiency and heat balance.

6.2.2 Polymers

The charged polymers have excellent lubricating properties compared with other polymers when water are used as solvent. As for these charged polymers, an ultra-low CoF less than 0.0006 is obtained. Another the important effect arises from the hydration layers surrounding each charged segment on the ionized polymers, which are formed many hydration sheaths.

The majority of these systems is aqueous-based, and most also require the tribochemical transformations of the surfaces during a running-in period before achieving superlubricity. Under high pressure and low temperature, an orthorhombic polymorph forms, leading to an anomalous dimple formation and a relatively high friction. Conversely, under low pressure and high temperature, a hexagonal polymorph emerges, exhibits a robust macroscale superlubricity. The lamellar structure of the hexagonal polymorph facilitates and interlayer sliding, is similar to the two dimensional materials used as solid lubricants.

Polyalkylene glycol (PAG) is generally used as synthetic lubricants in industries and copolymers manufactured via a combination of ethylene oxide and propylene oxide. The molecular-level hydrated top layer and a suitable amount of free water molecules are considered as the main reason for the observed superlubricity of CoF as low as 0.002 in the full immersion state. In the superlubricity system, superior load-bearing capacity is achieved by PAG aqueous solutions wherein multilayered adsorption layers by the hydrated PAG molecules on the sliding solid surfaces. The synergetic effect of the sufficient adsorption of molecules and the unique shear rheology of the PAG aqueous solution are essential to achieve superlubricity. The superlubricity mechanism for a neutral brush involves the permeation pressure of the solvent causing the polymer chains to spread throughout the aqueous solution, leading to the formation of molecular brushes and separation of the contact surfaces, which greatly reduce the friction. A solvent layer with low shear resistance between the brush layers helps to achieve superlubricity. The superlubricity mechanism of a charged brush is similar to that of a neutral brush at low pressure. Similar to polymers, macroscale superlubricity is achieved with hydrated alkali metal ions (Li^+, Na^+, K^+) based on hydration effect. This superlubricity is attributed to the hydration layer formed by alkali metal ions, providing the hydration repulsive force and having a liquid-like response to shear.

6.2.3 Acid-Based Aqueous Lubricants

CoF about 0. 004 is achieved betweena glass plate and a Si_3N_4 ball with the lubrication of phosphoric acid solution ($pH = 1.5$) after a short running-in period. Gradual vaparation of free water from the solution leads to a reduction of proportion of water molecules. When the proportion of water molecules reduces to a constant, the lowest CoF appeares. A stable hydrogen bond network is formed between phosphoric acid molecules and water molecules on two friction surfaces. The superlubricity mechanism of phosphoric acid is attributed to the hydrogen bond network formed between phosphoric acid and water molecules on the stern layer by the attached hydrogen ions.

Glycerol molecules contain three hydroxyl groups. The superlubricity of pure glycerol was observed at 80°C. The superlubricity is attributed to the easy sliding on triboformed OH-terminated surfaces. It is due to the triboinduced degradation of glycerol that produces a nanometer-thick film containing organic acids and water. This film can form a hydrogen bond network consisting of glycerol molecules adsorbed on the

OH-terminated surface by hydrogen bonds.

6.2.4 Oil-Based Lubricants

The superlubricity originated from TiNi alloy and steel under castor oil lubrication is obtained. The achievement of superlubricity hinges on two important factors. The first factor is the triboformed OH-terminated surface with nickel and iron oxy-hydroxide on the pin and steel in the contact region, respectively. The second factor is the positively charged surfaces induced by the hexanoic acid via the coordination and chemical reactions. It is the combination of the presence of Ni in the pin, iron in steel flat, and the branched OH group in the backbone of the castor oil molecules that permit the intercalated Fe/Ni intercalated nanostructures to be formed, resulting in ultimately ultralow friction.

6.2.5 Ion Liquid

IL is a series of well-known lubricants and lubricant additives in surface science and engineering, because they can improve the lifetime of the mechanical systems by reducing the friction and providing the wear protection. While studies on the superlubricity of IL have manily focused on low normal load (0.02 N) in combination with carbon quantum dots or at the microscale involving a SiO_2/graphite tribopair. Therefore, the macroscale superlubricity of IL under high normal loads is necessary. Researchers explore the robust superlubricity of IL aqueous solutions between a Si_3N_4/SiO_2 tribopair, achieving a CoF as low as 0.002 under neutral conditions ($pH \approx 6.9 \pm 0.1$).

6.2.6 Nanomaterial-Based Lubricants

The effectiveness of the nanodiamonds glycerol colloidal solution is investigated as a lubricant in glycerol solution. Although the stablė superlubricity (0.006) is achieved with both solutions after a running-in period, the colloidal solution exhibits significantly less wear. The superlubricity system based on nanodiamonds glycerol colloidal solution is discovered. The ultra low CoF is attributed to the hydrodynamic effect and the hydrogen bond layer.

Nanoparticles are significantly enhanced the study of oil-based liquid superlubricity. Only 1% of inorganic fullerene-like WS_2 nanoparticles dispersed in polyalphaolefin (PAO), an ultra low friction and wear is achieved under boundary lubrication conditions. The superlubricity of WS_2 sheets, which are formed on the rubbing surfaces and the combined rolling/sliding effects of the particles. Superlubricity with PAO by adding nano boron nitride (h-BN) and it is mainly attributed to the ball bearing effect and weak Van der Waals interaction force between nano boron nitride molecules at contact surface. When lubricated with the colloidal solution, the wear scar on the ball is much smaller than that of glycerol solution, indicating a substantial reduction in wear during the running-in period largely reduced. The ultra low CoF is attributed to the hydrodynamic effect and the hydrogenbond layer. The reduce in wear volume derives from the rolling effect of the nanodiamonds. Introducing nanopartides into the study of water-based liquid superlubricity provides a potential way to discover liquid superlubricity systems with good lubrication properties. Liquid superlubricity at the macroscale is focused on the mechanism, such as the silica layers and hydrodynamic lubrication for water, the tribochemical layer, and hydrogen bond network for viscous lubricants, and the stern layer and hydrogen bond network for acid-based lubricants.

6.3 Solid Superlubricity

Superlubricity with the existence of a nearly zero friction state and incommensurability is recognized to be the main mechanism for the realization of solid superlubricity.

6.3.1 Macro, Micro, and Nanosuperlubricity

In 1993, CoF in the range of 0.001 was obtained with a molybdenum disulfide coating in ultrahigh vacuum, which is attributed to the frictional anisotropy on the sulfur-rich basal planes during intercrystallite slip. The ultra low friction observed between graphite layers is arributed to their incommensurability. The interlayer shear stress in the superlubricity regime between two-layer graphene at the macroscale, and the residual stress, resulting in the random incommensurate stacking and the mismatch of the crystal lattice, is considered responsible for the facile shearing. Weak interatomic interactions between rotated layers and Coulomb repulsion are considered to play a part in the realization of ultra low friction, the main mechanism of solid superlubricity is generally accepted to be the incommensurability of the crystal planes. Robust superlubricity, without directional limitation, is observed under an ultra high contact pressure of 1 GPa with graphene and h-BN, which is due to the randomly oriented graphene nanograins at the multi-asperity contact area and, thus, induced overall incommensurability. Meanwhile, graphene nanoflakes are transferred onto the AFM silicon tip, achieving robust superlubricity with a CoF as low as 0.0003 between graphene (tip)/graphite, while no orientation dependency. This superlubricity can be maintained under an ultra high contact pressure of 2.52 GPa, which demonstrates the applicability to general engineering. Furthermore, superlubricity with a CoF as low as 0.0007 is observed between a graphite nanoflake-wrapped tip and highly oriented pyrolytic graphite.

6.3.2 DLC Films

The nanomaterials can be employed as solid lubricants when combined with DLC films in a dry nitrogen environment under rolling/sliding contact. Superlubricity is achieved through the formation of a carbon-rich superlubricious tribolayer at the interface, reducing the overall friction by a minimum of 20 times, and with no surface damage. DLC superlubricity often needs to be realized in inert gas. A sulfur-doped hydrogenated DLC film maintains an ultra low friction coefficient of 0.004 in an environment with 50% relative humidity due to the high binding energy of the S-C.

6.3.3 MoS_2, Graphite, and CN_x

Graphite is commonly used as a good solid lubricant in practical applications. It is composed of layers of graphene sheets and exhibits high strength, stiffness, and thermal conductivity along the basal plane. Its interlayers are bonded by weak Van der Waals forces, resulting in low shear strength. Therefore, graphite can be easily cleaved to obtain an atomically smooth surface, making it an ideal material for studying superlubricity. One method involves repetitively rubbing the spherical tip against the graphite surface and transfers the graphene nanoflakes to the surface of the spherical tip during the rubbing process. The frictional force between the graphene and the graphite surface is very small because of the weak shear strength and incommensurate contact.

Some solid lubricants, such as graphene, MoS_2, WS_2, black phosphorus, and diamond-like carbon films are considered as potential superlubricity materials. Achieving superlubricity involves the following two factors: (1) constructing an incommensurate friction interface and (2) reducing the frictional contact area of the two kinematic friction pairs. The graphitic-like/MoS_2 films with hierarchical structure are synthesized, which easily achieves macro superlubricity (0.004) under humid air. Solid lubricants include graphene/graphite and members of the transition metal dichalcogenides (TMDs) family, which are lamellar crystals with the general formula MX_2M being a transition metal (Mo, W) and X being a chalcogen (S, Se, Te). The weak inter-layer Van der Waals and electrostatic interactions allow layers to slide with relatively small effort. Namely: ①atoms at the sliding interface experience only weak interacting forces (e.g., Van der Waals), ② sliding surfaces are free from any contaminants and atomically flat, ③there is a non-zero misfit angle between the layers. There is a growing interest in understanding the mechanical and tribological properties of MXenes, however, no report of MXene superlubricity in a solid lubrication process at the macroscale has been presented until now.

6.3.4 Solid-Liquid Superlubricity

A robust superlubricity with CoF of approximately 0.004 is achieved by combining glycerol with the polymer coating. The superlubricity mechanism is attributed to the formation of a tribofilm, mainly composed of graphene nanoflakes in the contact zone. The extremely low friction achieved on the hydrophobic graphene coating with liquid can aid in the development of a high-performing lubrication system for industrial applications.

Derived from the method aiming to solve the insufficient load-bearing capacity issue of liquid superlubricity, a new category named solid-liquid combined superlubricity has emerged, which refers to the means of achieving superlubricity by combining liquid lubricants with solid additives. The superlubricity behavior of Si_3N_4/SiO_2 friction pairs using sodium hydroxide-modified few-layer black phosphorus (BP-OH) as a water-based lubricant additive is found. BP-OH nanosheets are formed on sliding surfaces to prevent direct contact of the asperity peaks, and the dynamically stable water is retained near the BP-OH nanosheet surface through hydrogen bonding, thus achieving robust superlubricity under a wide range of contact pressures (up to 1 GPa). It protects the substrates from chemical reactions. The low shear resistance between the layered nanosheets can significantly reduce the friction force when the asperity peaks make contact, and these solid nanosheets can bear a much high load than pure liquid superlubricity. A special water layer may develop on the surface of adsorbed nanosheets and further promote the performance.

6.3.5 Superlubricity Engineering

Rolling bearings and gears are critical components in mechanical systems such as automotive engines and gearboxes, wind turbine drive trains, and satellite solar-array-drive systems. DLC coatings and two-dimensional (2D) materials for rolling/sliding contact applications in an oil-free environment are revolutionized. The extraordinary material properties of graphene and other 2D materials are very attractive due to their potential for utilization as solid lubricants. The large-scale application of superlubricity, including applications in transportation, fluid machinery, manufacturing, space technologies, microelectromechanical systems, and renewable energy, may result in trillions of dollars of economic benefits annually worldwide. For the purpose of making the era of "superlubricitive engineering" a reality, massive efforts focusing on certain problems have been made, and the significant success thus far provides good preparation for the new era. However, key challenges are addressed for successful superlubricity applications: ①durability in complex practical working conditions. Further efforts are made to develop superlubricity technologies that can withstand extreme conditions like harsh high/low temperatures, ultrahigh/low speeds, and strong radiation, to provide new solutions for high-tech sectors such as aviation and aerospace. ②enhancing high-performance superlubricity systems. At present, the realization of an ultrahigh load-bearing capacity or ultralow friction is limited to certain materials and conditions, and it is necessary to develop more types of high-performance superlubricity systems and meet the industrial requirements under different conditions. ③further development of oil-based superlubricity. As one of the most important industrial lubricants, lubricating oils are also indispensable in the establishment of superlubricitive engineering. ④ cost-effective expansion from the nano/

microscale to the macroscale. In particular, the economic cost in the process of realizing superlubricity is taken into consideration. These identified challenges and gaps in the current research on superlubricity also create a roadmap for future development in this area. The arrival of "superlubricitive engineering" can be predicted, since the former achievements have laid a substantial foundation, and the enormous value in economic benefits and energy savings has raised sharply growing attention worldwide.

6.4 High Temperature Superlubricity

Many components in turbines, transportation manufacturing, and aerospace are operated at high temperatures. There is an urgent need for the lubrication solutions at high temperature in industrial applications, but high temperature superlubricity can be very challenging.

6.4.1 Design of materials

DLC is a common material in the field of tribology, with many researchers have studied its lubrication ability at high temperature. However, due to its poor thermal stability, achieving the ultralow friction behaviors of DLC films becomes increasingly difficult under high temperature conditions. The lowest stable CoF (0.008) of DLC films is observed at 600 ℃. The ultralow friction mechanism is a synergistic effect of the shielding effect of hydrogen at the contact surface and the repulsive electrostatic force among the self-generated oxide composites on the contact surface. High-temperature superlubricity on a nickel high-temperature alloy substrate using antimony trioxide (Sb_2O_3) and magnesium silicate hydroxide coated with carbon is found. The amorphous carbon film generated by the friction chemistry and the release of—OH and Si – O active groups contributes to the realization of high-temperature superlubricity. Hexagonal boron nitride is widely used in many fields for its high thermal conductivity and excellent chemical properties. The h-BN coating on the steel surfaces exhibits high-temperature superlubricity at 800 ℃, facilitated by the formation of oxide composites through tribochemistry.

6.4.2 Architecture of High Temperature Lubrication

As an ideal sealing material, graphite has self-lubricating properties, high thermal conductivity, excellent chemical stability, and corrosion resistance. It is worth noting that at high temperature, the oxidation resistance of graphite in atmospheric conditions is relatively poor, and leads to changes in its friction and wear due to the presence of oxidized abrasive particles. At present, in the research field of high-temperature superlubricity, numerous new superlubricity systems, are under development. Nonetheless, high-temperature lubrication technology with a CoF lower than 0.01 is still in its infancy.

6.5 Super Low Wear

The industrial revolution demands the introduction of high performance engineering materials to minimize energy loss, while also being environmentally friendly and economically effective.

6.5.1 Polymer

Ultra-low-wear polyethylene (ULWPE) manifests the best tribological performance compared with the most widely used artificial joint materials. Its high hardness and strength lays a solid foundation to a low wear volume, and its high ductility and hardness enable it to endure abrasive and adhesive wear, resulting in excellent wear resistance. Additionally, the high hardness, strength, ductility and good wetting of ULWPE materials reduce the damage of material to adhesion and abrasive wear, resulting in excellent wear resistance. The environmental dependence of ultra-low wear behavior of PTFE and alumina composites suggests tribochemical mechanisms. The tribological properties of materials represent a comprehensive behavior, which is involving multi-scales and affected by multiple factors including mechanical surface, physical-chemical interactions, structure, and boundaries.

6.5.2 2D Material

The urgent demand for atomically thin, superlubricating, and super-wear-resistant materials in MEMS systems has stimulated the research of friction-reducing and antiwear materials. Newly emerging two-dimensional (2D) materials of MXenes possess lots of merits, which provide the potential solutions for the lubrication issues in harsh conditions. Macroscale and atomicscale characterizations are utilized to explore the lubrication behaviors of the composite coatings and clarify the influence of the coating composition and tribo-test parameters in the establishment of ultra-wear-resistant sliding interfaces. The MXene/nanodiamond coating exhibits almost no wear when rubbed against PTFE ball. A nanostructured tribofilm with unprecedented bonding features is in situ formed along the sliding interface. The ultra-wear resistance highly depends on the combined effects of PTFE shielding and self-lubrication, MXenes layer shearing, and nanodiamond self-rolling nanodiamond. overall, two-dimensional MXene can effectively improve the wear resistance of the polymer matrix and has the potential to be used in friction-protective layers under complex working conditions.

6.5.3 Composite

Superior wear resistance and low friction in hybrid ultrathin silicon nitride/carbon films stem from the synergy of the interfacial chemistry and the carbon microstructure.

The superior performance of the hybrid film is attributed to the constructive synergy of microstructure and an enhanced interfacial chemistry arising from the additional interfacial bonding. This special engineering structure makes it possible for the synthesis of super-hard and super-durable lubricative coating for industrial applications.

6.5.4 Transfer Film and Link to Low Wear

Transfer film, a protective barrier that forms when a solid lubricant slides against a hard and high-surface-energy counterface, plays an important role in friction and wear reduction. Transfer films formed during the initial operation of the parent solid lubricant are promptly removed by ploymer pin. However, transfer films formed that develop after the solid lubricant has transitioned to ultralow wear rates themselves in the ultralow range, ranging from 10^{-8} to 10^{-10} $mm^3/N \cdot m$. Solid lubricant polymers generally have lower moduli and surface energies than opposing counterface materials, which are often ferrous. As the contact moves, adherent debris is left behind to be formed what is termed the transfer film. The transfer film protects the parent polymer from the counterface on subsequent contacts and therefore reduces the wear rate of the polymer. Fillers that successfully reduce friction and wear of polymers also produce thin and uniform transfer films. As a result, the wear-reducing effect of fillers in polymers is often attributed to their ability to improve transfer film thickness, uniformity, or adhesion. Naturally, the ability of the transfer film to protect the polymer is limited by its ability to adhere to the counterface.

The low-surface-energy solid lubricant transfers to the higher-surface-energy counterface to minimize energy. This is the reason why the steel probe removes the transfer film so effectively. The transfer film must remain stable to prevent the intermittent exposure of the high-surface-energy counterface. The higher the surface energy of the counterface, the lower the strength of the solid lubricant, and the smaller the debris, the more adherent the transfer film will be. When the PTFE composite lays down its own transfer film, the two surfaces are perfectly compatible and there is no single weak interface. The stability of a weak interface can involve the material properties, surface texture, operating conditions, and even environmental constituents. For probes with high surface energy, the weak interface is located within the film, at the film-counterface interface, or a combination of both.

6.5.5 Tribochemical Mechanism

Composites of PTFE and alpha phase alumina produce wear rates nearly five orders of magnitude lower than those of pure PTFE. The wear rate of composites is dependent on the humidity of the environment, suggesting a tribochemical mechanism contributes to their ultralow wear behavior. The ultralow wear rates are accompanied by tribochemical changes that stabilize surfaces by anchoring polymer chains to dispersed nanoparticles and the countersurface. By contrast, tribochemistry is extremely sensitive to the changes in wear rate regardless of the filler. By removing the protective effects of transfer films, the indexed reciprocation increases wear rates by 100-fold while effectively eliminating tribochemical accumulation and tribofilm formation on the polymer pin. The protective transfer films initiate low wear rates, facilitating tribochemical accumulation, which in turn reduces wear rates in a virtuous cycle.

6.5.6 Quantifying Properties of Transfer Films

Significant reductions in wear are accompanied by small wear debris and thin, more complete, and seemingly adherent transfer films. Improved transfer films contribute to reduce wear by shielding the composite from the hard counterface asperities. The cause-effect relationship between wear rate, debris size, and transfer film quality remains an open question and is likely system-specific. The observations of varied wear rates have corresponded to systematic changes in the appearances and physical characteristics of transfer films. The ubiquitous trend linking the reduced wear to improve transfer film quality has motivated strong suspicions that the improved transfer films contribute to reduce wear by protecting the solid lubricant from the inherently damaging counterface. The wear reduction mechanism may be due to more traditional reinforcement mechanisms such as mechanical reinforcement, preferential load support, crack arresting, and energy dissipation. Transfer film quality appears to be improved when debris are small. The chemical changes initiated via sliding create direct bonds between the polymer/filler and polymer/counterface, which stabilizes the near surface of each while bonding the transfer film to the counterface. Ultra-low-wear sliding is permitted when the polymer and transfer film are sufficiently dissimilar to establish a weak sliding interface between them. These ultralow wear rates cannot persist without stable and persistent transfer films. Bulk polymers simply deposit a new transfer film and reestablish ultra-low-wear sliding. Transfer films are failed because they lack sufficient material to establish a new equilibrium.

6.6 Superlubricity Mechanism

Superlubricity is generally classified into solid superlubricity and liquid superlubricity according to the lubricants present at the interfaces. Superlubricity describes a phenomenon where the friction force between two sliding surfaces nearly vanishes ideally or a state in which the sliding CoF is less than 0.01 in actual mechanical lubricating systems.

6.6.1 Solid Superlubricity

Solid superlubricity at the nanoscale or microscale is easily achieved with two-dimensional materials, such as molybdenum disulfide, graphite flakes, boron nitride, and graphene. It has also been achieved at dissimilar interfaces, including the crystalline gold/graphite interface, graphene nanoribbons/gold interface, and silica/graphite interface.

Solid superlubricity lubricants are mainly based on incommensurate contact and weak interfacial interaction between layers. Most solid lubricants exhibiting superlubricity possess a laminate structure, which enables weak interaction between layers and can also realize the incommensurate contact, providing favorable conditions for superlubricity. However, superlubricity is closely dependent on test conditions and the environments. So far, no single solid lubricant has been found to achieve superlubricity in all environments. Some kinds of solid lubricants, such as graphite, require humid environments or other gases with high concentration to achieve superlubricity.

6.6.2 Liquid Superlubricity

Liquid mechanisms are typically divided into three categories: hydrodynamic effect, electric double layer effect, and hydration effect. In some cases, superlubricity can be obtained merely through one mechanism, while in other cases, combinations of two or three mechanisms are necessary.

Numerous liquid lubricants exhibit macroscale superlubricity behavior, including water, oils, and polymers. The superlubricity mechanisms of these lubricants account for the formation of hydrogen bond networks, tribofilms, hydration layers, or molecular brushes, depending on the properties of the lubricants. Compared to solid lubricants, liquid lubricants are less constrained by environment conditions. It is easy to found that all of the present superlubricity liquids are water-based, due to the very low coefficient of viscous pressure of water. The other liquid superlubricants such as ceramic materials with water, phosphoric acid solution, and glycerol solution with acid or polyhydric alcohol, show the potential applications, because all of them can achieve superlubricity on the traditional tribometers. The hydration model requires the surface to be charged, the tribochemical model requires the surface reacting with water, and the hydrogen bond network requires the surface to be OH-terminated. There are several methods to meet the requirements of liquid superlubricity. The first method is to choose the surface on which the tribochemical reaction can occur, such as ceramic. The second one is to treat the surface by grafting molecular brushes, self-assembled monolayers, or changing the hydrophobicity. The third method is to choose the soft surface with high content of water.

6.6.3 Superlubricity at Atomic Scale

Researchers observe superlubricity in scanning tunneling microscope (STM) measurements when a monocrystalline tungsten tip slides on a Si (001) surface. The superlubricity seems to be retained up at high speed. The experimental observations of the structural superlubricity have also been reported on MoS_2 and Ti_3SiC_2. The application of a few volts across the piezoelement shaking the probing tip is sufficient to enter the dynamic superlubricity regime. With the suitable parameters, the reduction in friction can also lead to a reduction of wear.

6.6.4 Simulations of Superlubricity

The developments in both hardware and methodologies have allowed the computer simulations to play an important role in the interpreting experiments and understanding the origins of energy dissipation during nanoscale sliding. The interesting phenomenon is superlubricity, characterized by dry sliding with very low kinetic friction that has been observed in experiments of incommensurate surfaces in the absence of adsorbates. The phenomenon of friction vanishing also arises in theoretical models, consisting of one-dimensional infinite atomic chains such as the Frenkel-Kontorova-Tomlinson model. In this model, rigid incommensurate surfaces can exhibit zero-static friction with nonvanishing surface energy corrugation and zero-kinetic friction in the zero-velocity limit. Note that zero-static friction always implies zero-kinetic friction although zero-kinetic friction is not necessarily imply zero-static friction. Zero-kinetic friction is achievable only under the condition of one stable state.

There are many problems to be solved as follows: ① is there a uniformed mechanism of the superlubricity? ② what kind of the molecular structure can realize superlubricity? ③ is there a relationship between nanoscale superlubricity and macroscale superlubricity? ④ can the superlubricity be obtained from the oil-based liquids? The superlubricity can be applied to the practical engineering, offering potential energy savings during the friction process and contributing to reduceing the environmental pollution.

6.7 Superlubricity System

Given the fact that approximately one third of global energy consumption is caused by friction, achieving superlubricity, which means the friction nearly vanishes and the CoF is less than 0.01, and thought to be one of the most effective approaches to address this issue.

6.7.1 2D Material

Two-dimensional (2D) materials with a layered structure are excellent candidates in the field of lubrication due to their unique physical and chemical properties, including weak interlayer interaction and large specific surface area. For the last few decades, graphene has received lots of attention due to its excellent properties. Besides graphene, various 2D materials (including MoS_2, WS_2, WSe_2, $NbSe_2$, $NbTe_2$, ReS_2, TaS_2, and h-BN) have been found to exhibit a low CoF at the macro- and even micro-scales, which may lead to the widespread applications in lubrication and anti-wear. A relatively new, large, and quickly growing family of two-dimensional early transition metal carbides and nitrides (MXenes) also present a great potential in different applications.

6.7.2 Superlubricity System

Klein's group obtained the earliest progress in liquid superlubricity by preparing grafted polymer brushes on mica surfaces, achieving CoF on the order of 0.001 or less. Kato performed the remarkable works on liquid superlubricity with the ceramic/water systems, but the main challenge still lies in the narrow contact pressure range, which is always less than 20 MPa. The superlubricity with a CoF of 0.004 between glass/Si_3N_4 friction pairs are found under the lubrication of a phosphoric acid aqueous solution (pH=1.5), which becomes the starting point of phosphate-based superlubricity. Various superlubricity systems have since been developed, induding the biomaterial superlubricity systems, acid mixed solution-based superlubricity systems, oil-based superlubricity systems, and alkaline solution-based superlubricity systems.

6.8 Brief Summary

This chapter provides an introduction to superlubricity, highlighting its significant advantages in industrial applications and energy savings. The liquid superlubricity, solid superlubricity, and high temperature superlubricity are introduced separately.

7 Bionics Tribology

7.1 Introduction

The movement patterns of organisms have undergone complex environmental effects of hundreds of millions of years of screening for the survival of the fittest. Current species and their movement patterns are the result of optimization under the existing natural conditions. The structural features and mechanisms of the action of organisms in nature will be a treasure trove of rich inspiration for bionic research, bionic engineering design, and development.

As an interdisciplinary field, the research of bionics involves many areas, for instance science, medicine, mechanics, materials science, physics, and chemistry. The structure of the research object has expanded from the macro scale to the micro scale, even to the nano scale. Great inventions of the modern era such as flying machines, submarines, robots, etc. have been inspired by natural systems. There are two methods of biomimetics used in engineering design processes: biomimetics by analogy and biomimetics by induction. Biomimetics arises from the quest to address a problem in a design by trying to draw analogies from natural systems. Biomimetics by induction begin from basic scientific approaches in biology with no immediate intention to apply them in the engineering systems. The field of friction, wear, and lubrication, although relatively new, has also drawn numerous inspiration from nature. For instance, there are crawling of insects, locomotion in reptiles, lubrication in eyelids, wear and lubrication in bone joints etc.

7.2 Definition and Fundaments

The Institute of Bionic Tribology is concerned with how to get inspiration from the biological structures and apply them in engineering application practices. The tribological research of bionic applied the biological system generally takes morphology as its main objective is to achieve the bionic transformation and application.

7.2.1 Nature Inspired Friction and Wear

Living species are chosen to undergo natural selection, which is actually the survival of the fittest. Some typical examples are provided: ① snakes have scales that can provide hight friction, which facilitates their movements, ② the fascicle tip of a female mosquito can penetrate human skin without pain, ③the setae on the pad of gecko strongly adhere to the substrate, ④ the lotus leaf is super-hydrophobic, which contributes to its self-cleaning, ⑤ the scales of pangolin have remarkable wear resistance against the soil particles, ⑥aloe vera mucilage serves as a natural lubricant. Typical researche and applications of bionic tribology are shown in Fig. 7 - 1.

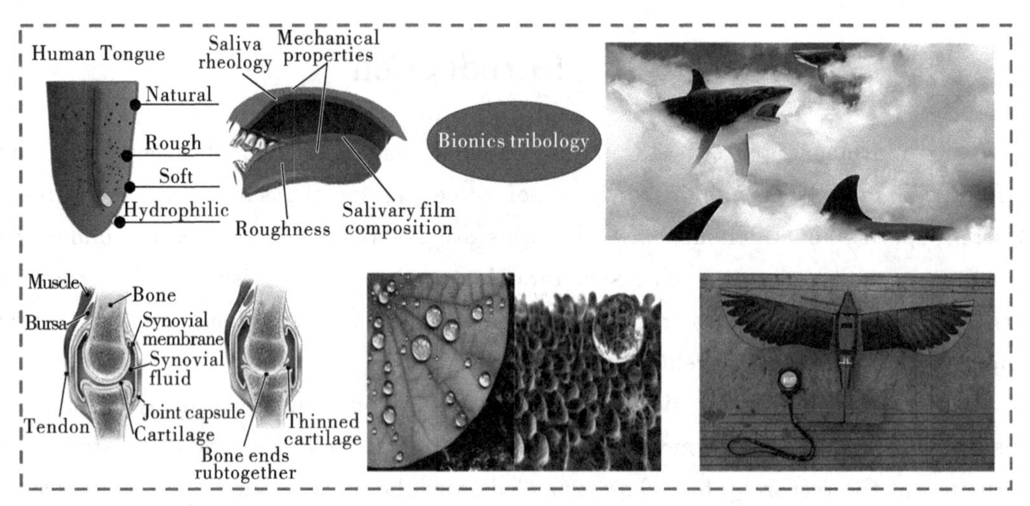

Fig. 7 - 1 Research and application of bionic tribology

CoF varies in distinct relative sliding directions, which is usually called the friction anisotropy. When concertina-climbing on the inclined surfaces, snakes tilt the ventral scales relative to the ventral surface and interlock with the asperities on the rough surfaces, which enhances friction during climbing. Some beetles possess distal hairy adhesive pads combined with small claws to ensure a high adhesion force on a smooth substrate, approximately 36 times their own body weights while the size of setae greatly affects the friction and adhesion on the rough substrates. The claws at the end of legs are usually used to enhance the friction on rough surfaces by employing interlocking.

7.2.2 Nature Inspired Lubrication

Lubrication is a common reason for the fast movement in various environments for some creatures. This property could help the creatures reduce the energy barriers for moving, escape from predators, and catch up with or trap prey. Human eyes and joints have been widely observed and studied on the context of superlubrication, so the lubrication abilities not possessed by human beings are discussed.

Fish can move very fast and efficiently underwater. The mucus of fish, which is a water solution of the secretions produced by goblet mucous cells, is the key lubricant to effectively relieve the water resistance. When a fish is swimming fast in the water, due to the compositions in the mucus, such as water-soluble macromolecules, the transition of the boundary flow from laminar flow to turbulent flow is significantly suppressed which greatly reduces the drag force. The hydrophilic skin mucus of loach and eel not only plays a dominant role in drag reduction but also the has the antifouling properties in an aqueous environment to freely swim in sludge or water.

For example, aloe vera mucilage with a viscosity of approximately 5.8 mP·s at room temperature is regarded as a new potential environmental-friendly lubricant. Brasenia schreberi mucilage is another new type of lubricant such that CoF of a glass surface could be as low as 0.005, which is a state of superlubricity. This property is connected with its nanosheet structure and water molecules bonded on the polysaccharide nanosheets which participate in the formation of hydration. The above-mentioned lubricating phenomena are mainly controlled by the function and characteristics of special liquid materials. The skin of the shark utilizes directionally the grooved wedge-shaped scales on the epidermis to achieve excellent drag-reduction properties. This special structure is widely recognized to reduce the turbulence intensity with the back-flow induced by the sloped groove and scale arrangement.

7.2.3 Future Prospects

These excellent tribological properties are the results of evolution. The biological lubricant and fascicle tip of a mosquito would inspire more breakthroughs in the medical field such as a painless syringe needle. The role of the biological systems in tribology is indispensable. Different natural tribological systems are served as an inspiration for the various developments in the field of tribology. Followings are various areas where a lot of developments are still required which can serve as potential: ① the use and development of lubricants based on the articular cartilage joints for reducing friction and wear, ②development of the composite materials based on the teeth of the mammals for better antiwear and antifriction properties, ③development of the clutch materials based on the attachment pads of the gecko to create better adhesive forces, ④developments of the biolubricants based on natural lubricants for better biodegradability, ⑤development of the extreme pressure lubricants based on the lubricant composition of the mammalian joints for sliding contacts under high loading conditions.

7.3 Bio-Tribology

The word "biotribology" is first used and defined by Dowson in 1970 as "those aspects of tribology concerned with biological systems."

7.3.1 Biological Attachment and Detachment

The attachment and detachment of the organism and the external environment is also an important research area. Adhesion usually refers to the resistant force encountered when two contacting surfaces are separated vertically. The fast movement of a gecko on a vertical or inverted wall of trees or buildings, the strong attachment of an octopus or a diatom to the substrate underwater, and the super hydrophobic surface of a lotus leaf demonstrate adhesion. Research on the mechanism of gecko attachment and detachment has shown that changes in the contact angle between the spatula-like end of the plantar bristles and the base correspond to adhesion changes. When the contact angle is greater than 30°, the contact surface begins to break. The factor that affects the mechanism has been proven to be Van der Waals force considering that humidity could enhance the Van der Waals force. Capillary formation requires a high humidity of over 50%-60% RH. The Van der Waals force between a spatula and a substrate in dry conditions can be estimated.

Tree frogs living in the tropical rainforests or humid environments have special structures on their toe tips. Their survival activities are mainly on trees, and they can adhere to slippery substrates during crawling. A tree frog utilizes wet adhesion, which is strongly dependent on the capillary force. The surface at the distal end of a tree frog's pad has hexagonal flat-topped epithelial cells separated by fluid-filled grooves, where the fluid can be squeezed out or sucked in according to the relationship between the width of grooves and the surface separation. Mussels also provide strong adhesive force underwater with the byssus attaching to almost any type of inorganic and organic surfaces, including polymers.

7.3.2 Natural Synovial Joints

Tribological studies of natural synovial joints and joint replacements are usually carried out in terms of friction, wear, and lubrication, with the specific considerations of the biological nature of the joint system. The natural synovial joints are examples of excellent tribological systems. The bearing surfaces are articular cartilage, while the lubricant is synovial fluid. Articular cartilage is generally described as biphasic, consisting of both solid and fluid phases. These unique biphasic properties are important for the lubrication mechanisms in the synovial joints. Synovial joints have a large range of motion, and different types of joints enable various actions. Although joints have various forms and complexity depending on their positions and animal species, they typically have basic structures such as articular surface, joint capsule, and joint cavity. The dynamic mechanical environments of the load and speed are also important to consider when examining the tribological mechanisms.

7.3.3 Total Joint Replacement

Artificial joints are used to treat joint diseases such as osteoarthritis and trauma. Friction, lubrication, and wear studies of artificial joints are important to optimize the performance of these man-made bearings and improve clinical functions. Currently, the major clinical challenge associated with artificial joints is the loosening of prosthetic components, mainly due to wear debris-induced adverse biological reactions. Wear debris is largely generated at the articulating surfaces as well as fixation interfaces. In addition, metallic ions may be released as a result of corrosion. Furthermore, artificial joints are implanted in patients by surgeons, and therefore patient-specific variations in anatomy, loading/motion patterns, and surgical techniques significantly affect how the joint implant functions in the body. It is important that the tribological studies are integrated with the biological and clinical studies on artificial joints.

7.3.4 Synthetic Cartilage and Lubricant

Various biomaterials have been developed in different combinations to reduce wear and wear debris generation in artificial joints. Currently, the majority of artificial joints utilize a material combination of UHMWPE against cobalt-chromium counterface for both the hip and knee joints. CoF is the lowest, while the hydrogel/hydrogel has the highest CoF. The contact between hydrogel on hydrogel causes great deformation during friction.

To improve the tribological properties of hydrogels, most of the early researchers used organic boundary lubrication to functionalize hydrogels, which leads to the formation of an interfacial water film due to the hydrophobicity of the hydrocarbon chain. The main wear mechanism of rubber is abrasion wear, while boundary lubrication functionalized hydrogel is adhesive wear.

7.3.5 Soft Tribology

Soft surfaces, especially those belonging to the category of hydrogels, are often characterized by low frictional behaviors with CoF, as low as 10^{-3}. These low friction coefficients have been attributed to loose hydrophilic polymer chains on the hydrogel surface, which can create a soft, hydrated polymer layer by entrapping water. Adhesion and separation is often present simultaneously, which makes the frictional behaviors of soft systems complex to measure, analyze, compare, interpret and understand.

Movements of several groups of muscles in the tongue result in variations in the stiffness of the organ. In the presence of saliva, the tongue exhibits more hydrophilic behavior due to the adherence of amphiphilic proteins present in saliva. In lubricated deformable polyacrylamide hydrogels, it is found to cause a decrease in CoF with the increasing normal forces. The ratio between the stiffness of two interacting surfaces such as the relatively hard palate and the soft tongue, may therefore also be considered. When pairing hard surfaces with soft surfaces, deformation may occur, leading to changes in the surface structure. For spherical contacts, Hertzian type deformation can be expected depending on the Young's moduli of the surfaces in contact.

7.3.6 Skin

Skin is the largest organ of the human body. It covers 1.6 to 2.0 m^2 of the surface area in adults and accounts for approximately 16% of a person's weight. In daily life, human skin comes into contact with a variety of materials due to labor, exercise, maintaining warmth, health and beauty needs, which cause many skin friction problems.

Human skin is a multilayered composite material composed of the epidermis, dermis, and subcutis. It is a soft biomaterial with a complex anatomical structure and has a complex material behaviour in the mechanical contacts with objects and surfaces. It exhibits highly non-homogeneous, nonlinear elastic, anisotropic, viscoelastic material properties similar to those of soft elastomers. Under dry skin conditions, adhesion caused by the attractive surface forces at the skin material interface, as well as deformation (hysteresis, ploughing) of the soft, viscoelastic bulk skin tissue, contribute to CoF. Adhesion is considered the main contributor to the friction of human skin, whereas deformation mechanisms are assumed to play a minor role. Skin friction depends on the type (solid, soft, or fibrous material) and physical properties of the contacting materials, as well as on the physiological skin conditions (e.g., hydration state, sebum level) and the mechanical contact parameters, especially the normal load, i.e., contact pressure. In addition, sliding velocity, age, gender, ethnicity and anatomical region are also influencing factors.

7.3.7 Hair and Textile Fibers

Everybody hopes to have smooth and soft hair, similar to skin. Grooming and maintaining beautiful hair is a daily process for most people, with friction and adhesion being the most relevant parameters in hair care. CoF is useful as it serves as a quantitative marker for human perception of texture and current investigations focus on the correlating friction properties at macro- and micro-scale with hair structure under various experimental conditions. All these investigations have been used to provide a formulation guide of hair-care products and treatments.

7.3.8 Oral Tribology

Oral tribology concerns all aspects of tribology related to the oral systems. Human oral cavity is composed of the palate, chin, teeth, tongue, mucosa, and glands. The temporomandibular joint (TMJ) connects the palate to the chin. Friction and wear in the mouth is normally related to the processing of various foods, oral hygiene, and orthodontics, and thus it is unavoidable. Generally, oral tribology involves the studies of teeth, saliva, TMJ, and the soft tissues of the oral cavity. Four kinds of test methods have been used to simulate the tribological behaviors of oral system: clinical investigations in vivo, in vitro testing, in situ testing, and finite element analysis. Tooth wear, either natural or artificial, mainly results from mastication. Mastication is the action of chewing food, which involves an open phase and a closed phase. No occlusal forces are involved in the open phase, resulting in little or no tooth wear, while during the closed phase, the occlusal load is applied to the foods and the hard particles in foods are dragged across the opposing surfaces, causing occlusal surface wear.

7.3.9 Ocular Tribology

The eye is a typical example of a lubricated moving system in the human body. The tear film acts as a protective layer for the cornea and maintains the optical smoothness for vision. Ocular tribology has attracted much attention, particularly since the introduction of contact lenses. There are two major tribological interactions in the eye. One interface occurs between the eyelid and cornea in the natural eye, while in the case of contact lenses, two interfaces are formed between the contact lens, the eyelid, and the cornea. Major tribological issues have focused on the properties of the tear film and its lubrication mechanism, adhesion, and friction of various contact lenses, and adhesive comfort to develop a satisfactory contact lens.

7.3.10 Micro Circulation

Tribological problems in the cardiovascular system mainly involve artificial heart valves, intravascular stents, as well as capillary blood flow. Safety and service lifespan have been emphasized both from the mechanical aspect due to wear damage and from the pathological aspect due to the penetration of wear particles into blood vessels. In the human circulatory system, red blood cells undergo large deformation while passing through capillaries with diameters ranging from 4 to 8 μm. Understanding the tribological behaviors, particularly the flow properties of capillary blood flow are important both for protecting red blood cells from damages during microcirculation and for providing clinical guidance. A number of investigations have been carried out, including experimental simulation in vitro, and corresponding theoretical analysis on the behavior of blood flow in microvessels and the motion of red blood cells in capillary tubes.

7.3.11 Key Issues in Biotribology

The major issue currently associated with the bearing surfaces of artificial joints is wear and subsequent wear debris which can cause adverse tissue reactions and the loosening of the prosthetic components. One of the key lubrication mechanisms associated with biphasic articular cartilage is the fluid load support due to its fluid phase. A general theoretical analysis is presented on the fluid load support and the corresponding friction coefficients. It is now generally accepted that fluid inside articular cartilage is squeezed out under load and consequently lubricates the cartilage surface. The fluid squeezed out is a function of the loading time and decreases as the loading time increases. This is consistent with the friction-loading time relationship observed experimentally. The boundary lubrication mechanism of articular cartilage is equally important. A number of lubricating constituents have been identified, including high viscosity hyaluronic acid, glycoproteins, and lipids.

7.3.12 Preclinical Tribology

Biotribology deals with the application of tribological principles, such as friction, wear, and lubrication between relatively moving surfaces in medical and biological systems. Biotribology plays an important role in a number of medical devices.

Dental restorations include materials used to restore the function, integrity, and morphology of missing tooth structure, as well as the replacement of missing tooth structure that is supported by dental implants. Excessive wear could result in the failure of dental restorations and implants. For gastrointestinal diseases, passing a gastrointestinal endoscope through the digestive tract to the lesion location, and conducting operations such as checking, ablation, removing or stripping the diseased tissue, are the most basic clinical treatments for the digestive system. During gastrointestinal endoscopy, the endoscope is pushed into the human digestive tract with the aid of outside force, which may cause a series of complications such as throat abrasion, bleeding, mucosal tearing, and perforation of the digestive tract due to repeated insertion, rotation, pushing, and retrieval operation.

7.4 Development of Friction Elements

7.4.1 Solid Materials

Metals and alloys are mainly applied to the orthodontic appliances and dental implants nowadays, with only amalgam filling is used for dental restoration. Amalgam, commonly called amalgam alloy, is composed of a mixture of mercury and powdered alloy made mostly of silver, tin, copper, etc. and has been successfully used as one of the most popular direct restorative materials. The major attraction of amalgam alloy is its proven longevity due to its high wear-resistance from the metallic character in clinical service and ease of clinical use.

Some high-toughness dental ceramics have been developed in the last decades, aiming to minimize the damage by brittle fractures. Yttria-stabilized tetragonal zirconia polycrystal, a high-toughness zirconia ceramic, has been increasingly accepted to overcome the issues and can be used as an alternative to porcelains or glass-ceramics in posterior restorations.

7.4.2 Liquid Materials

Lubrication between the contact lens and the eye is critically important. Fluid film lubrication in the presence of the tear film ensures minimum friction, smooth motion, and negligible damage to the eye during blinking. At the back of the contact lens, which is in contact with the ocular surface, the speed is relatively low and the brush lubrication mechanism is also dominant. Important considerations include the lens materials, wetting agents, as well as the tear film, and particularly their interactions. The structure of the tear film and composition (proteins, lipids, and mucin) are critically important in the tribological and clinical functions of a contact lens. Soft contact lens materials with high water content surfaces or the incorporated wetting agents, such as poly-vinyl-pyrrolidone (PVP) or poly-vinyl-alcohol (PVA) have been developed to reduce friction between the contact lens surface and the lid-wiper.

7.4.3 Natural Materials

Various biomaterial combinations are used for the articulating surfaces of artificial joints. These are broadly divided into soft-on-hard and hard-on-hard combinations. Soft-on-hard combinations mainly include UHMWPE against cobalt-chromium alloys or alumina/zirconia toughened alumina composite ceramics. Titanium alloys are sometimes preferred, particularly for total disc replacements in the spine, due to their low elastic moduli and improved imaging quality, but surface treatments to improve its wear-resistance are necessary. Hard-on-hard bearing surface combinations for a hip implants include metal-on-metal, ceramic-on-ceramic, and ceramic-on-metal. The major source of wear debris in soft-on-hard combinations is from the UHMWPE bearing surface. Polyether-ether-ketone (PEEK) has been extensively investigated as a potential material to replace UHMWPE, in particular against a ceramic counterface for the hip, the knee, and the spinal disc. In addition, PEEK-on-PEEK combinations have been investigated for small joints such as the spinal disc. More recently, the UHMWPE-on-PEEK bearing combinations are suggested as a candidate for knee implants.

7.5 Manufacture of Bionic Tribology Surface

7.5.1 Layer by Layer Assembly

The idea of layer by layer assembly comes from the structure of fish scales. The fish scales of carassius auratus includes 3 parts: a basal, an apical, and two laterals. The basal is in the cutis of the fish body, and the laterals are covered by surrounding scales, while the apical is the only part subjected to the friction between the fish body and water.

7.5.2 Lithography Freeze-Casting

Freeze casting is a promising method in this field to duplicate the architecture of natural materials with relatively high similarity. It mainly utilizes growing ice as a template to make materials form a lamellar microstructure in the space between the ice crystals by controlling the freezing conditions to build various architectures with nanomaterials. Furthermore, the multiple factors are controlled to regulate local structural characteristics, such as size, amount of porosity, wall thickness, interlamellar bridging, and roughness of internal surfaces, and further controlling local properties through altering slurry concentration, freezing rate, sintering, temperature gradient, etc.

7.5.3 Templating Additive Manufacturing

Besides freeze casting, additive manufacturing also duplicate various structures flexibly. Due to the unique way of additive manufacturing to build architectures through stacking materials, the local properties are regulated efficiently through spatially controlling local microstructures and compositions from the bottom up in a layer-by-layer sequence. Further, abundant techniques give additive manufacturing a high chance to fabricate the advanced bioinspired materials, such as direct ink writing, 3D magnetic/electrical printing, slip-casting, etc. Bouligand structures are duplicated through electrically assisted 3D printing. Two parallel electrodes are used to control the aligning direction of multi-walled carbon nanotubes (MWCNT) accurately and then the stage rotates with a fixed angle to stack the next layer. Besides, the biomimetic materials like brick-and-mortar are also be fabricated through magnetically assisted slip-casting.

7.5.4 Spray and Spin Method Self-assembly

Additionally, self-assembly is also the common method to mimic the architectures of natural materials. Self-assembly commonly associates and gathers individual components together to form the close and orderly architectures through physical or chemical processes, such as electrophoretic deposition, filtration, and evaporation. The abiotic tooth enamel with remarkable mechanical properties is fabricated by hydrothermal growth of ZnO nanowires layer by layer. In addition, the bouligand structure mentioned above is also be fabricated through in-situ polymerization with cellulose nano-crystals.

7.5.5 Electroplating Shortage

The factors are considered in the process of designing and fabricating bionic materials, such as interface, composition, and gradient, to work together and achieve remarkable enhancement for the mechanical properties. On the other hand, traditional bionic manufacturing technologies often ignore the subtle structures in the interface, such as the bridges in the interface of nacre, which leads to bionic materials being far from able to attain the enhancement effects like natural materials. Besides, unequal scale replication of structures in artificial materials also restricts the improvement of the mechanical properties due owing to the lack of toughening on a microscale and even nanoscale level. Hence, the development of materials inspired by living organisms puts forward a high requirements of manufacturing technology to imitate and even the transcend nature.

7.5.6 Bionic Self-repairing Friction Coating

Self-healing materials have drawn great attentions recently owing to their ability to automatically restore the monphological integrity and anticipated properties after being damaged, inspired by the fitness-enhancing functionality of living organisms. There are two strategies for obtaining self-healing functionality: molecular engineering for intrinsic self-healing materials and the delivery of external healing chemicals for extrinsic self-healing materials. Many intrinsic self-healing materials with the ability to heal damage multiple times in situ are designed and synthesized based on dynamic covalent chemistry and supramolecular chemistry. Research on self-healing materials provides methods to maintain or even completely recover the properties of materials under conditions of cracks, punctures, and corrosion. The development of a dual functional intelligent composite or coating with self-healing and self-lubricating functions at the same time offers a new route towards safer and longer-lasting products. Several self-healing agents, such as tung oil, linseed oil, and hexamethylene diisocyanate with lubricating abilities, have been microencapsulated successfully for intelligent lubricating materials with self-healing functionality.

7.6 Bionic Lubrication and Lubrication Systems

7.6.1 Solid Materials

Metallic materials are widely used as biomedical materials for the substitution of hard tissues, scaffolds in regenerative medicine, implants, medical devices, and so on. The first generation of biometallic materials is stainless steel. The second generation consists of Co-Cr-based alloys, which are widely used in artificial hip and knee joints because of their wear resistance ability. The third generation comprises titanium-based alloys, which are extensively studied and used in the implantation of artificial bones because of their good biocompatibility, mechanical properties, and wear resistance ability. Another biometallic material is magnesium and its alloys, which are used as biodegradable materials. Because of the low elastic moduli of magnesium and its alloys, they are used to replace materials for damafed bones. Significant researches have been conducted on zirconia, including the development of stabilized zirconia, induction of bioactivity by chemical treatments, and improvement of surface hardness.

7.6.2 Intelligent Lubricating Materials

Dry/humid adaptive lubricating materials are essential in high vacuum or aerospace equipment, which also are exposed to moisture before launch. Transition-metal dichalcogenide (TMD) is excellent lubricants in vacuum but lose properties in humid air due to oxidation. DLC, by contrast, exhibits low friction in humid environments. The WC/DLC/WS_2 coating grown by the magnetron-assisted pulsed laser deposition has demonstrated a hard, humidity-adaptive nanocomposite with the ability of lubrication recovery in dry/humid environment cyclings.

7.6.3 Natural Materials

Metallic and ceramic materials have been broadly applied in biomedical engineering in the form of implants and devices because of their excellent mechanical performances such as high moduli and stiffness. Ceramic materials also have some disadvantages such as low fracture toughness. Therefore, scientists have conducted abundant researches on the tribology of hydrogel.

7.7 Tribological Properties of the Bionic Tribology System

7.7.1 Adherence at Fluid-Solid Intersurface

Saliva is the most important component of the chemistry of the human mouth. It consists of approximately 98% water and a variety of electrolytes and proteins. Salivary proteins are selectively adsorbed onto all solid substrata as well as mucosal membranes exposed to the oral environment and then formed an acquired salivary pellicle within seconds.

An important function of saliva is to form a boundary lubrication system and serve as the lubricant between hard (tooth) and soft (mucosal) tissues to decrease the wear of teeth, and reduce the friction of oral mucosa and tongue surfaces, thereby preventing lesions and making swallowing easily. The salivary film is formed by a layer-by-layer adsorption of salivary proteins, resulting in a heterogeneous structure consisting of a thin and dense inner layer and a thick, highly hydrated, and viscoelastic outer layer. The inner layer of the salivary film effectively decrease the friction and tooth wear, thus playing an important role in the lubricating properties of saliva.

7.7.2 Adherence to Self-Cleaning of Solid Surface

Adhesion is also be avoided in some cases. The lotus leaf is commonly known as the representative of super-hydrophobic and self-cleaning surfaces because contaminants on the leaf are removed by rolling water droplets. This characteristic is ascribed to the hierarchical structures of a lotus leaf, with a microstructure of papillose epidermal cells covered with 3D epicuticular wax tubules. When a water droplet is on the surface, the tiny cavities of the surface are filled with air, which leads to a contact angle of over 160° and very weak adhesion between the droplet and surface. The colocasia leaf, contrary to the lotus leaf, with the papillose protuberance on its surface separated with webbed ridges, also exhibits this super-hydrophobic property. This structure-induced super-hydrophobicity is usually explained by the Wenzel-Cassie model. Similar hydrophobic properties also exist in some animals. The water strider can walk freely on the water surface because of its legs with hierarchical hairy structure.

7.7.3 Friction between Solid-Fluid Interfaces

Lubrication and cleaning are inseparable, and lubricants themselves can even cause environmental fouling. Lubricating materials with porous structures and infused liquids provide a slippery surface that is able to be hydrophobic, and exhibiting the potential to develop intelligent self-lubricating materials with self-cleaning/antifouling functionality. To mimic the lubricating capabilities of earthworms, lubricants are stored as discrete droplets in the supramolecular polymer matrix with a textured structure on the surface. The liquids are site-specifically and quickly released/self-replenished under external mechanical stimuli. The released oils are stabilized by the surface texture to form thick lubricating layers, reducing friction and enhancing wear resistance. Moreover, the coatings exhibit excellent antifouling properties in a sticky soil environment, which is attributed to the slippery hydrophobic lubricating layer on the surfaces.

7.7.4 Wear Characteristics of Bio-Inspired Anti-Wear Technology

The degree of wear resistance determines the service life of products or natural components, especially for creatures living in sea and soil environments with continuous impacts from water flows or abrasion from sand particles. Some creatures gain excellent wear resistance by controlling the flow field during friction with specific surface textures, such as the mole cricket living underground and frequently burrowing soil. The soil-engaging component of the mole cricket has evolved into a structure with high wear resistance. The top and side surfaces of its tergum are covered with long and short setae, forming a hierarchical morphology combined with thorns. During digging, the friction properties of the tergum surface are good in the direction of setae growth compared with other directions. Another typical soil-living animal with high wear resistance is the pangolin, which experiences wear from micro-plowing, micro-shoveling, and stress fatigue.

7.8 Applications and Developments

7.8.1 Synovial Joint Lubrication

In recent years, the prevalence of people with amputations has risen due to production safety accidents, traffic accidents, diseases, aging, and other factors. Artificial limbs enable amputees to retain upright mobility capabilities and restore appearance. The suspension system and socket fitting of artificial limbs play major roles and have vital effects on the comfort, mobility, and satisfaction of amputees. Coupling between the prosthesis and the transtibial stump is typically achieved by a socket, which is a critical component for the prosthetic performance and the sole means of load transfer between the prosthesis and the stump in current prosthetic practice.

7.8.2 Gecko

The dry adhesion of the gecko setae leads to the creation of surfaces with strong and reversible adhesive properties, which are used to fabricate wall-climbing robots. For synthetic dry adhesives to perform well on rough surfaces and engage a high percentage of the microfeatures across large patch sizes, a hierarchical suspension structure is required. The hierarchy may, first, help the contacting features align and conform to the surface, and second, distribute the applied load to the contacting elements evenly.

7.8.3 Spider Silk Fiber

To overcome the 'last kilometer' problem for the complete removal of multiple heavy metal ions from water, it is important to solve the imbalance of the amphiprotic functional groups and address the low total group density of amphoteric cellulose-based adsorbents while maintaining the swelling of the fiber structure. Inspired by the principle of spider silk, a novel amphoteric bionic fiber adsorbent is designed via electrostatic assembly of cellulose nanofibers (CNF) as the skeleton, and graphene oxide (GO) and polyethyleneimine (PEI) as the outer and inner layers, respectively.

7.8.4 Friction-Reducing Shark Skin

Skin of the shark utilizes directionally grooved wedge-shaped scales on the epidermis to achieve excellent drag-reduction properties. This special structure is widely recognized for reducing the turbulence intensity with the backflow induced by the sloped groove and scale arrangement. The viscous drag reduction caused by the surface profile of shark skin is comparable to that of the smooth surfaces of Gray's paradox, which has great streamlining of the body and special behavior. Shark skin has also revolutionized swimwear to help athletes obtain higher speeds.

To date, there are several theories or hypotheses for the drag reduction of shark skin. One is protruding height theory, i. e., the tips of micro-grooves stick out the viscous sublayer to efficiently decrease the intensity of turbulence. Another is attack angles of placoid scales. The third is the mucus on shark skin. Under the effect of fluid, the mucus can develop in nano-long chains extending into the viscous sublayer, which can raise the depth of the viscous sublayer playing as a buffering layer and reducing the turbulent intensity.

7.8.5 Bio-Mimicked Lubrication

To reduce friction while being non-poisonous, non-fungoid, and meeting other medical requirements, a bionic medical lubricant is synthesized. There are natural lubricants in the human digestive system, to help the canal motion evacuate waste. Artificial synovial fluids already are used in bone joints. The tribological properties and medical feasibility of using eel fish synovial fluid as a reference have been studied. The hydrogels are designed to mimic the tribological properties exhibited by the mucus on the fish skin. Extensive research has been carried out on artificial hydrogels to achieve CoF of the order 10^{-4}, which is lesser than natural fish skin. The gel friction during lubrication is found to be due to the hydrated layer of polymer chains when they are non-adherent or repulsive from the substrate and the friction due to elastic deformation is shown to be caused by the absorbed polymer chain.

7.8.6 Air Lubrication of Emperor Penguins

Emperor penguins at the antarctic poseess an extraordinary skill to reduce hydrodynamic drag when accelerating inside water or jumping out of water onto slippery ice sheets. Air lubrication is the method of injecting air into the boundary layers which are also a technique that has been used to propel ships and torpedoes at high speeds through sea water. This high speed during ascent is attributed to the penguins emitting air bubble clouds into the turbulent boundary layer over most of their body surface. This is attributed to the plumage of penguins which are water repellent due to peen oil being present on them unlike in marine applications where it is still a difficult task to maintain sufficient bubble coverage within the turbulent boundary layer.

7.9 Brief Summary

This chapter introduces bionics tribology, biotribology, and the development of bio-inspired drag reduction applied in daily life and industrial production.

8 Green Tribology

8.1 Introduction

Green tribology is defined as the science and technology of the tribological aspects of ecological balance and environmental and biological impacts. The specific field of green or environmental-friendly tribology emphasizes the aspects of the interacting surfaces in relative motion which are important for energy or environmental sustainability or have an impact on today's environment. The main idea of green tribology is shown in Fig. 8 - 1.

Green tribology can be understood in the broader context of two other 'green' areas: green engineering and green chemistry. Green engineering is defined as the design, commercialization, and use of processes and products that are technically and economically feasible while minimizing: ① the generation of pollution at the source, ②risk to human health and the environment. The three tiers of green engineering assessment in design involve: ① process research and development, ② conceptual/ preliminary design, ③the detailed design, including pollution prevention, process heat/ energy integration, and process mass integration. Another related area is green chemistry, also known as sustainable chemistry, which is defined as the design of chemical products and processes that reduce or eliminate the use or generation of hazardous substances.

There are 12 principles of green tribology: the minimization of ①friction, ②wear, ③the reduction or complete elimination of lubrication, including self-lubrication, ④ natural, ⑤biodegradable lubrication, ⑥ using sustainable chemistry and engineering principles, ⑦biomimetic approaches, ⑧ surface texturing, ⑨ environmental implications of coatings, ⑩real-time monitoring, ⑪ design for degradation, ⑫ sustainable energy applications. We further define three areas of green tribology: ① biomimetics for tribological applications, ②the environmental-friendly lubrication, ③ the tribology of renewable energy applications. The integration of these areas remains a primary challenge for this novel area of research. We also discuss the challenges of green tribology and future directions of research.

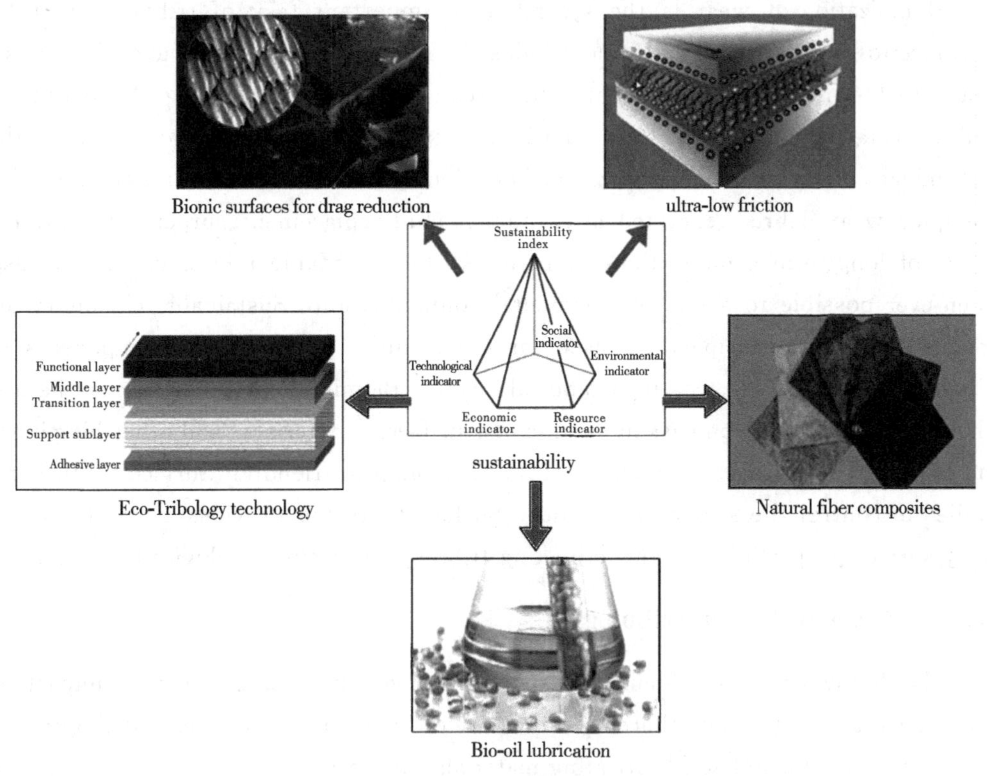

Fig.8 - 1 Core ideas of green tribology

8.2 Green Tribology

Since tribology is an interdisciplinary area that involves, among other fields, chemical engineering and materials science, the principles of green chemistry are applied to green tribology as well. However, since tribology involves, besides the chemistry of surfaces, other aspects related to the mechanics and physics of surfaces, there is a need to modify these principles.

8.2.1 Principles of Green Tribology

Friction is the primary source of energy dissipation. According to some estimates, about one-third of the energy consumption is spent to overcome friction. Most energy dissipated by friction is converted into heat and leads to the heat pollution of the atmosphere and the environment. The control of friction and friction minimization, which leads to both energy conservation and prevention of damage to the environment due to heat pollution, is a primary task of tribology. It is recognized that for certain tribological applications (e. g., car brakes and clutches), high friction is required; however, ways of the effective use of energy for these applications should be sought as well.

Minimization of wear is the second most important task of tribology that has relevance to green tribology. In most industrial applications, wear is undesirable as it limits the lifetime of the components and poses challenges for recycling. Wear can also lead to catastrophic failure. In addition, wear creates debris and particles that contaminate the environment and can be hazardous for humans in certain situations. For example, wear debris generated after human joint-replacement surgery is a primary source of long-term complications in patients. Biodegradable lubrication is also used when ever possible to avoid environmental contamination. Sustainable chemistry and green engineering principles are used for the manufacturing of new components for tribological applications, coatings, and lubricants. Biomimetic approaches may be used whenever possible, encompassing biomimetic surfaces, materials, and other biomimetic and bioinspired approaches as they are more ecological-friendly. Surface texturing is applied to control the surface properties. Surface texturing provides a way to control many surface properties relevant to making tribo-systems more ecologically friendly.

8.2.2 Areas of Green Tribology

The following three focus areas of tribology have the greatest impact on environmental issues, and, therefore, they are of importance for green tribology:

(1)Biomimetic and self-lubricating materials/surfaces.

(2)Biodegradable and environmentally-friendly lubricants.

(3)Tribology of renewable and/or sustainable sources of energy.

Biomimetics is the application of the biological methods and systems found in nature to the study and design of engineering systems and modern technology. Biomimetic materials are also usually environmental-friendly in a natural way since they are a natural part of the ecosystem.

(1) The Lotus-effect-based non-adhesive surfaces. The term 'Lotus effect' stands for surface-roughness-induced Superhydrophobicity and Self-cleaning. Superhydrophobicity is defined as the ability to have a large (more than 150°) water contact angle and, at the same time, low contact angle hysteresis. The lotus flower is famous for its ability to emerge clean from dirty water and repel water from its leaves. This is due to a special structure of the leaf surface (multi-scale roughness) combined with hydrophobic coatings. These surfaces have been fabricated in the laboratory with comparable performance. Adhesion is a general term for several types of attractive forces that act between solid surfaces, including the Van der Waals force, electrostatic force, chemical bonding, and the capillary force, owing to the condensation of water at the surface. Adhesion is a relatively short-range force, and its effect (which is often undesirable) is significant for microsystems that have contacting surfaces. The adhesion force strongly affects friction, mechanical contact, and tribological performance of such system surfaces, leading, for example, to "stiction", which precludes microelectromechanical switches

and actuators from the proper functioning. It is therefore desirable to produce non-adhesive surfaces.

(2)The Gecko effect, which refers to the ability of specially structured hierarchical surfaces to exhibit controlled adhesion. Geckos are known for their ability to climb vertical walls due to strong adhesion between their toes and various surfaces. They can also detach easily from a surface when needed. This is due to the complex hierarchical structure of the gecko foot surface. The Gecko effect is used for applications when strong adhesion is needed. The Gecko effect is combined with self-cleaning abilities.

In the area of environmentally-friendly and biodegradable lubrication, several ideas have been suggested; The use of natural (e.g. vegetable-oil-based or animal-fat-based) biodegradable lubricants. This involves oils are used for engines, hydraulic applications, and metal-cutting applications. In particular, corn, soybean, and coconut oils have been used so far (the latter is of particular interest in tropical countries such as India). These lubricants are potentially biodegradable, although in some cases, chemical modification or additives for best performance are required. Vegetable oils can exhibit excellent lubricity, far superior than mineral oil. In addition, they have a very high viscosity index and high flash/fire points.

Powder lubricants particularly, boric acid lubricants, are generally much more ecologically frciendly than traditional liquid lubricants. Boric acid and MoS_2 powder are used as an additive to the natural oil. Friction and wear experiments shows that the nanoscale (20 nm) particle boric acid additive lubricants significantly outperform all of the other lubricants for the friction and wear performance. The nanoscale boric acid powder-based lubricants exhibit a wear rate more than an order of magnitude lower than the MoS_2 and larger-sized boric acid additive-based lubricants.

The tribology of renewable sources of energy is a relatively new field. The following issues can be mentioned.

(1) Wind-power turbines have a number of specific problems related to their tribology and constitute a well-established area of tribological research. These issues include water contamination, electric arcing on generator bearings, problems related to the wear of the mainshaft, gearbox bearings, and gears, the erosion of blades (solid particles, cavitation, rain, hail stones), etc.

(2) Tidal-power turbines are another important way of producing renewable energy, which involves certain tribiological problems. There are several specific tribological issues related to tidal-power turbines, such as their lubrication (seawater, oils, and greases), erosion, corrosion, and biofouling, as well as the complex interactions among these damage modes. Besides tidal power, the ocean-water flow energy and river flow energy (without dams) are used with the application of special turbines, which provide the same direction of rotation independent of the direction of the current flow.

8.2.3 Friction Induced Self-Organization

Physically, friction and wear are related to the second law of thermodynamics, since they are the results of the tendencies of energy to dissipate and material to deteriorate. Friction and wear are the complex phenomena that do not have single mechanisms but rather involve the physical and chemical processes of different natures. While friction and wear are irreversible processes that normally lead to material deterioration, under certain circumstances, friction can lead to self-organization or increase orderliness when various 'secondary structures' (e.g., in situ formed tribofilms) are formed at the frictional interface. Self-lubrication is an example of such friction-induced self-organization.

Friction-induced self-organization has a significant potential for the development of self-lubricating, self-healing, and self-cleaning materials, which can reduce environmental contamination, making it relevant to green tribology. The reduction of friction and wear due to the self-organization at the sliding friction interface can lead to self-lubrication, i.e., the ability to sustain low friction and wear without the external supply of a lubricant. Since lubricants often pose environmental hazards, while friction and wear often lead to heat and chemical contamination to the environment, self-lubrication has the potential for green tribology. Self-lubrication is also common in the nature, making it of interest to scientists and engineers exploring biomimetic approach, which also has the potential for green tribology.

The concept of the 'selective transfer' or the 'nondeterioration effect' is developed. The selective transfer is a type of friction characterized by the formation at the contact interface of a thin nonoxidized metallic film with a low shear resistance that is not able to accumulate dislocations. According to the selective transfer scheme, a selective dissolution of a component of a copper alloy occurs, followed by the transfer of the component to a contacting body (a steel shaft). The standard example of selective transfer is the formation of a copper layer in a bronze-steel system lubricated by glycerin.

8.2.4 Self-Lubrication

Protective coatings is one of the best material solutions for the above subjects. There are three directions to make the optimum design of the protective coatings: coatings of solid lubricants, hard but lubricious carbon-based coatings, and in-situ formation of tribofilms. In the first direction, MoS_2 is directly coated or mixed with the metallic elements onto the tool surface. Since this coating works as a solid lubricant, low wearing, and friction state is preserved as long as MoS_2 is left in the system. Second, diamond-like-carbon (DLC) containing metallic elements as well as diamond coatings are favored for dry machining. In particular, normal pressure is relaxed by W-containing soft DLC. In the case of Si-containing DLC, the friction is significantly reduced by the tribofilm, which is synthesized by in situ polymerization. This third approach is further exploited to develop more reliable tribofilms.

Self-lubrication via in situ formation of tribofilms. At the first stage of the low-temperature, tribochemical surface reaction, free oxygen atoms are penetrated into titanium nitride films. This self-lubrication process is directly applied to the reduction of wear and friction in various titanium alloy automotive and electrical parts and components. Since titanium alloys are easy to cause adhesive wear against ceramics and metallic alloy counterparts, this type of mechanism is effective in putting low friction and wear state into practice for dry metal forming, dry power transmission, or dry moving transfer. Especially, this technology finds various applications in micro-machines and robotics working in dry environments.

8.2.5 Self healing

One of the most remarkable properties of the biological surfaces is their ability to self-heal and repair the damage caused by friction and wear. Self-healing is in apparent contradiction with the second law of thermodynamics, which states that dissipative processes, such as friction, are irreversible. However, in many nonlinear thermodynamically open systems that operate far from equilibrium, including the frictional systems, self-organization can occur. It is therefore important to investigate the thermodynamics of friction and self-organization during friction in order to design biomimetic self-healing surfaces. Friction and wear are complex phenomena that involve many physical and chemical processes, such as deformation, adhesion, abrasion, fracture, ploughing, chemical reactions, capillary condensation, and many others.

8.3 Eco-Friendly Lubrication

Lubricants are in high demand in industrial and automotive processes, as they minimize surface wear and extend the lifespan of the mechanical components. Currently, crude oil and its derivatives are of fundamental importance to the global economy and serve as the main source of lubricants. However, the presence of combustion residues (e.g., iron, calcium, magnesium, and zinc particles) of mineral oil and its disposal generate pollution in aquatic and terrestrial ecosystems, which determines low biodegradability. These oils contaminate soil, air, and water and strongly affect animal and plants.

8.3.1 History of Eco-Tribology

The "eco" in eco-tribology means not ecological but economical at this time. The climate changed by greenhouse effect comes to be admitted in a scientific manner, and the Kyoto Protocol was adopted in 1997. The proportion of ecology in ecotribology rise more than economy against the background of the rise of people's concerns to environmental problems. In addition, the global economic crisis in 2008 had a big influence for eco-tribology also. The role of an environmental performance has grown as an important factor deciding one product's value. Ecology and economy, which has been indicating the conflicted directions, becomes with the same meaning for eco-tribology now.

8.3.2 Solid-Liquid Compound Lubrication

Solid-liquid synergistic lubrication can compensate for the shortcomings when acting alone. In the solid-liquid composite lubrication system, the solid lubrication film can significantly improve the bearing capacity and lubricity of the friction surface by reducing the shear strength, thus acting as a spare lubricant, especially under the condition of boundary lubrication. And the liquid lubricant can further enhance the solid lubrication and reduce its wear. There is few relevant research on the synergistic effect of solid lubricating film and lubricating oil additives on friction response.

The use of solid lubricant coatings may also improve the tribological properties of sliding contact interfaces under boundary-lubricated sliding conditions. Sometimes, the beneficial synergistic effects of the combined uses of solid and liquid lubricants on friction and wear of sliding tribological interfaces can be expected. When fluid and boundary films fail, such coatings can carry the load and act as a back-up lubricant. Soft metals (such as In, Ag, Sn, Pb, Au) can provide reasonably low CoF to sliding surfaces under dry sliding conditions. When such coatings are subjected to sliding tests at elevated temperatures in the presence of synthetic oil, very low friction and wear can be achieved. In one circumstance, metal may catalytically decompose lubricants, for example, alumina in hard disc applications speeds up the degradation of perfluoropolyether oil. Therefore, an overcoat layer is required to limit direct reaction between lubricant and substrate, such as ceramic or CNx film on magnetic storage materials. It is also shown that the friction can be remarkably stabilized by plasma nitriding treatment of ALSI 52100 steel under lubrication conditions. An inert hard carbon film would be ideal overcoat by preventing direct contact between lubricating oil and metal substrate, meanwhile providing very low friction and wear to sliding tribological interfaces under dry sliding conditions. On the other hand, since the temperature and environmental parameters greatly influence the stability of friction and wear of DLC coatings, the use of lubricating oil will isolate the coatings from the surrounding hostile environment and act as a coolant to keep the temperature in the allowable limit. A combination of solid lubrication and liquid lubrication will find more and more successful examples and contribute to energy saving by lowering. CoF and by prolonging antiwear life.

8.3.3 Oil gas a Minimum Quantity Lubrication

Minimum quantity lubrication (MQL) is established on the idea that a drop of liquid is breached by an air stream, dispensed in streaks and transferred along the direction of airflow. This methodology of cooling has been developed to overcome the demerits of dry machining, so as to maintain ecological and economical balance leading towards product, process, environmental and economical sustainability. The various types of MQL system have been explored. The MQL phenomenon can be pertained in two different manners namely external and internal method of lubrication. The external MQL system in which the mixture of cutting fluid plus air possessing definite pressure is imparted through nozzle to various rubbing, shearing as well as the elevated temperature zones while machining. In addition, this gaseous medium has also been practiced these days making it more useful. On the other hand, in the later system coolant is not supplied via nozzle, but through the special designed tool. In MQL system, a tiny mass of vegetable oil or biodegradable synthetic ester is sprinkled to the tool tip with the compressed air. The oil consumed during machining operations is in the order of around 0.16~8.3 ml/min.

The cooling is achieved by air jet, whereas the lubrication occurs due to the fatty film of lubrication oil. Since the limitations of dry machining and wet cooling are not identical therefore the possible alternative to aforementioned machining conditions seems to be minimum quantity lubrication. Using this facility reduced quantity of pressurized jet of cutting fluid mingled with air is applied and regulated on different cutting area. Especially, MQL has been practiced for machining advanced material used in aerospace and bio-medical application because dry machining of same would not remain economical due to the elevated temperature, high friction and rapid wear. The interesting feature of MQL is the oil amalgamated jet of air targeted on various spots, resulting the cooling action due to hydrogen embrittlement action.

8.3.4 Multiscale Surface Texturing

From the Stribeck curve, one way to reduce friction is to shift transition of the lubrication mode from elastohydrodynamic lubrication to boundary lubrication for a high load and low speed. One direct way to achieve this is to create surface texture on oil lubricating surfaces which can be produced by machining, ion beam texturing, etching techniques and laser texturing. Among them, laser texturing is extremely fast, clean to the environment and provides excellent control of the shape and size of the micro-dimples for final optimum designs. Recently, studies have shown that laser surface texturing in the shape of dimples can offer transition from boundary lubrication to hydrodynamic lubrication.

The hydrodynamic lubrication region of textured surface is observed shift to the low speed and high load condition compared with non-textured original surface. The CoF under the boundary lubrication is also reduced on laser textured surfaces. The tribological properties of texture are normally influenced by the characteristics of dimple or grooves such as size, shape, orientation, depth and density. The width and the height dimples can be optimized to achieve low friction and wear at the sliding and rotating contact interfaces. Wang et al. investigated the influence of the micro-pits' density and size on the load-carrying capacity. Load carrying capacity can be increased 2.5 times higher for the textured than for the un-textured surfaces. Potential benefits of applying laser surface textured (LST) piston rings have been demonstrated theoretically and experimentally. A potential friction reduction of about 30% compared to non-textured rings under full lubrication conditions can be achieved. Moreover, surface depressions can act as a reservoir for lubricants, capable of feeding the lubricant directly between two contacting surfaces, and for trapping wear debris to reduce wear. This may prolong the endurance of the lubricant and minimize the amount of deformation in the sliding process.

Friction reduction by surface texturing has a wide variety of industrial applications including cylinder liner, piston ring, hard disk, sliding bearing, mechanical seal and even lip seal, etc. When texturing is used in surface treatment of the cylinder liner, oil consumption can be reduced by 70%, friction by 40% and wear by 80%.

8.3.5 DLC Coatings

DLC generically describes a class of amorphous carbon coatings in a manner analogous to the term steel describing a range of ferrous alloys. Carbon, as solid material can exist with different atomic arrangements and bonding (allotropes). The two most common allotropes are cubic diamond with tetrahedrally arranged sp^3 bonded carbon atoms and the hexagonal layered sp^2 bounded graphite. DLC is a family of metastable amorphous carbon consisting of a network of carbon atoms with a mixture of sp^3 and sp^2 bonds, exhibiting both the high hardness of diamond and low friction of graphite.

Currently, there are different types of DLC and efforts are on the way in many countries for standardization in their classification. There are three broad groupings of DLC coatings; hydrogenated, hydrogen-free, and doped/alloyed. Properties and performance of DLC coatings for the most part are guided by the sp^2/sp^3 bond ratios. A useful way of representing different types of currently available DLC is the ternary phase diagram of sp^2, sp^3 and H. Coatings containing a large fraction of sp^3 bound (70%) are usually designated Ta-C (tetrahedral amorphous), while coatings that are predominantly sp^2 are designated a-c. If the coating contains significant amount of hydrogen, they are classified as hydrogenated DLC and usually designated Ta-C:H or a-c:H depending on the ratio of sp^3/sp^2 bonds. Of course, without hydrogen, the coatings are classified as hydrogen-free. With or without H, DLC coatings are sometimes doped or alloyed with other elements. Non-metallic dopants such as Si, N, B, F have all been used. Metallic elements such as Ti, W, Cr, etc. have also been added DLC to create the so-called metal containing DLC. The primary purpose of doping and alloying is to control various desirable properties and enhance thermal stability of DLC coatings. Dopings and alloyings are used to increase adhesion, reduce residual stresses, improve the fracture toughness and retard the thermal decomposition of DLC to graphite at high temperatures.

In general, the properties of DLC coatings are determined by the hydrogen content, the doping element, if any, and the sp^3/sp^2 bond ratio. Some trends are observed for the properties of DLC coatings. For instance, coatings with high levels of sp^3 bond tend to have high hardness and elastic modulus, and hydrogenated DLC exhibits low hardness and modulus. DLC coatings are currently used for a variety of the tribological applications ranging from razor blades to aerospace components. The coatings are used to hard disk, automotive parts, gears, bearing and biomedical components.

8.4 Green Lubricants and Materials

A lubricant is a material used to facilitate relative motion of solid bodies by minimizing friction and wear between interacting surfaces. Examples of such applications include hydraulic fluids, electrical transformer fluids, heat transfer fluids and metal working coolants.

8.4.1 Vegetable Oil-Based Lubricants

The potential candidates for ecofriendly lubricants include vegetable oils, animal fats, and synthetic esters. Vegetable oils are the semisolid or liquid plant-derived composed of glycerides of fatty acids.

The polar nature of vegetable oils makes them fairly good solvents, which flush dirt and wear particles of metal surfaces. Vegetable oils show good boundary lubrication over mineral oils. Vegetable oils have positiveimpact on CO_2 level, considering that the released CO_2 in the atmosphere from their combustion is only a part (ranging from 30% to 50%) of that absorbed from plants during their growth. Strong adsorption of long-chain fatty acids present in vegetable oils to metallic surface assures high load carrying capacity of vegetable oils over mineral-based oils. Thermal properties of vegetable oils are improved due to the presence of a large number of carboxylate groups in the ester molecules and thereby a thick lubricating film can be maintained under highly loaded and high-slip contacts. High thermal conductivity increases the amount of energy transferred from the elasto-hydrodynamic contact to the surrounding material and fluid, thereby lowering the temperature in the contact zone and increasing viscosity. The high molecular weight of the triacylglycerol molecule and the narrow range of the viscosity change with temperature ensure the low volatile nature of vegetable oils. High linearity of the oils permits the triglycerides to maintain stronger intermolecular interactions with the increasing temperature than the branched hydrocarbons or esters. This ensures the high viscosity index of vegetable oils. Vegetable oil-based lubricants are more effective in reducing the emission levels of carbon monoxide and hydrocarbons. The high flash and fire points of vegetable oils than mineral-based oils and animal fats indicate their better fire resistance. Vegetable oils are readily and completely biodegradable into carbon dioxide and water molecules. Vegetable oils composed of fatty acids are generally nontoxic to the aquatic and terrestrial environments. The sources of vegetable oil are crops and they promote self-reliance as ample production capacity exists, but the source of mineral oil is a finite mineral deposit.

8.4.2 Ionic Liquids and Molecular Gel Lubricants

Ionic liquids (ILs) are defined as organic salts comprised entirely of ions, which exhibit melting points below 100 ℃. ILs have unique properties such as high chemical and thermal stability, almost negligible vapor pressure, non-flammability, high ionic conductivity and ease in dissolving organic, inorganic and polymeric materials. Besides, it is important to stress the possibility of tuning the ILs' physico-chemical properties such as melting point, viscosity, density, polarity and solubility through the combination of different cations and anions, making them suitable for many applications. The range of applications of ILs is very wide including the use as electrolytes in batteries or for metal electrodeposition, and as alternative solvents for the organic synthesis and catalysis, instead of the commonly used organic ones that are toxic and harmful to the environment. ILs are investigated as neat lubricants in several types of contacts and, when compared to traditional lubricants, such as perfluoropolyether and phosphazene, they exhibit superior tribological performance, namely the friction and wear reduction.

8.4.3 Extreme Pressure Additive

At high temperatures or under heavy loads where more severe sliding conditions exist, compounds called extreme pressure (EP) additives are required to reduce friction, control wear, and prevent severe surface damage. These materials function by chemically reacting with the sliding metal surfaces to form relatively oil insoluble surface films.

In addition to normal break-in wear, nascent metal (freshly formed, chemically reactive surfaces), time, and temperature are required to form protective surface films. After the films are formed, the relative motion is between the layers of surface films rather than the metals. The sliding process can lead to some film removal, but replacement by further chemical reaction is rapid so that the loss of metal is extremely low. This process gradually depletes the amount of EP additive available in the oil, although the rate of depletion is usually very slow. The severity of the sliding conditions dictates the reactivity of the EP additives required for maximum effectiveness. The optimum reactivity occurs when the additives minimize the adhesive or metallic wear without leading to appreciable corrosive or chemical wear. Additives that are too reactive lead to the formation of excessively thick surface films, which have less resistance to attrition, so some metals are lost by the sliding action. As a particular EP additive may have different reactivities with different metals, it is important to match the additive metal reactivity to the additives not only with the severity of the sliding system but also with the specific metals involved.

EP agents are usually compounds containing sulfur, phosphorus, chlorine, borate, or metals, either alone or in combination. The EP compounds used today depend on the end use application and the chemical activity that is required. Sulfur compounds, sometimes used in combination with chlorine or phosphorus, are used in many metal cutting fluids. Sulfur and phosphorus combinations are used in most industrial and automotive gear lubricants. These materials provide excellent protection against gear tooth scuffing and have the advantages of better oxidation stability, lower corrosivity, and often lower friction than other combinations that have been used in the past. The use of metallic EP additives is diminishing because of the influence of environmental concerns. Heavy metals are considered pollutants, and their presence is no longer welcomed in the environment. Based on performance needs and cost, ashless EP additives (dithiocarbamates, dithiophosphates, thiolesters, phosphorothioates, thiadiazoles, benzothiazoles, amine phosphates, phosphites, phosphates, etc.) may be preferred for some lubricant applications.

8.5 Smart Lubrication

8.5.1 Self-Healing Functionality

Intelligent lubricating materials containing liquid lubricants greatly improve the lubricating properties and prolong the lifetime by releasing lubricants at the friction interface. Self-healing materials have drawn great attention recently due to their abilities to automatically restore the morphological integrity and anticipated properties after being damaged, inspired by the fitness-enhancing functionality of living organisms. There are two strategies for obtaining the self-healing functionality: molecular engineering for intrinsic self-healing materials and the delivery of external healing chemicals for extrinsic self-healing materials. Many intrinsic self-healing materials with the ability to heal damage multiple times in situ are designed and synthesized on the basis of dynamic covalent chemistry and supramolecular chemistry. Capsules and vascular networks separate the active reagents from the matrix in the extrinsic self-healing materials, and heal cracks by the release of active reagents. Research on self-healing materials provide methods to maintain or even completely recover the properties of materials under conditions of cracks, punctures, and corrosion. The development of a dual-functional intelligent composite or coating with self-healing and self-lubricating functions at the same time offers a new route toward safe and long-lasting products. Three strategies for endowing intelligent lubricating materials with the self-healing function have been reported: the introduction of bifunctional active ingredients, the integration of both lubricating and healing components and the design of intrinsic bifunctional materials.

8.5.2 Self-Cleaning Functionality

Inspired by biological objects such as lotus leaves, cicada wings, snail shells, fish scales, shark skins, pitcher plants, and photosynthesis, self-cleaning materials have attracted much attention in industries and daily life because they can avoid frequent washing, hence the labor can be saved. The self-cleaning surface is primarily categorized into hydrophobic and hydrophilic surfaces. Hydrophobic materials clean themselves by sliding and rolling droplets, while hydrophilic materials use appropriate metal oxides to sheet the water and carry away dirt. Furthermore, oleophobic and amphiphobic surfaces are also promising self-cleaning materials.

Lubrication and cleaning are inseparable, and lubricants can even cause environmental fouling. Lubricating materials with porous structures and infused liquids provide a slippery surface that is hydrophobic and it exhibit the potential to develop intelligent self-lubricating materials with self-cleaning/antifouling functionality.

8.5.3 Self-Adaptive Functionality

Material selection is usually guided by a couple of factors including the operational humidity, temperature, geometry, load and speed, because different materials behave optimally under a limited range of the environmental conditions. Self-adaptive intelligent materials accept and respond to external information, and automatically change their own states to adapt to the environmental changes. It is termed for the ability to resist friction by changing surface chemistry and microstructures in different conditions, much like a chameleon automatically changing its skin color to best evade predators.

Dry/humid adaptive lubricating materials are essential in high vacuum or aerospace equipment, which are also exposed to moisture before launch. TMDs are excellent lubricants in vacuum but lose properties in humid air owing to oxidation. DLC, by contrast, exhibits low friction in humid environments. The WC/DLC/WS_2 coatings grown by the magnetron-assisted pulse laser deposition first demonstrate a hard humidity adaptive nanocomposite with the ability of lubrication recovery in dry/humid environmental cyclings. A graphite-like transfer film provides lubrication and seals the WS_2 phase from oxidation in humid air, while WS_2 is crystallized and re-oriented into hexagonal plated.

Broad temperature adaptive lubricating materials are developed to meet the need for high temperature engines. There are many lubricants that perform effective lubrication at medium or high temperatures, such as Ag, MoS_2, oxides, and fluorides.

8.5.4 Self-Reporting Functionality

Self-reporting materials, which indicate damage or stress with a clearly perceptible optical or electrical signal, have the potential to combine with intelligent lubricating materials and realize the real-time monitoring of wear or lubrication state. The triboelectric nanogenerator (TENG) initiated the field of nanoenergy was invented. TENG has been widely used as a self-powered sensor for various applications and offers a prospective solution to achieve real-time monitoring of wear states without external energy or software. The dynamic wear monitoring and positioning of the sliding bearing systems with a wear sensor array have been achieved successfully.

Triboluminescence (TL) directly converts the mechanical energy input from the outside world into light energy, which can not only supplement the demand for artificial generation of light/electric excitation sources but also have the potential to be used as a wear monitor in mechanical sensing and artificial intelligence fields. A self-lubricating polymer with a continuously self-regulating secretion system is produced, which is able to maintain a constant thin liquid overlayer, self-heal, and self-report its liquid content. Few intelligent lubricating materials are applied in mechanical components as the commercial materials, except for some solid lubricating coatings and composites with a wide-temperature range. There are still many technical challenges to overcome for transforming laboratory demonstrations into practical applications across a broad cross-section of industries.

8.6 Sustainable Tribology

8.6.1 Tribological Considerations

Archard's equation is a basic technique that is used often for predicting and analyzing wear life. The equation predicts that the wear volume is a linear function of sliding distance and load and is inversely proportional to the hardness of the material, which conforms to the general experience. For life-cycle assessment by standard wear tests, it is important to consider the following factors.

Identifying the predominant type of wear is a key factor for the selection of standard wear tests. A great deal of experience is generally required to analyze the surface topography created by wear and determine if wear is produced by a chance encounter or by recurring contact. Changes in the wear environment, such as the nature of the counterbody and load vector, can significantly alter the relative rankings of materials with respect to their wear rates and the life cycles. Standard wear test results of materials for the predominant wear type provide the comparative wear data of materials. Wear data on dimensional loss from a test surface in standard tests are used to assess the life span of materials for known dimensional tolerances. The wear rate in a steady state is useful in assessing the life cycle of the material. The duration of the standard tests may or may not yield wear data in the steady state.

Archard's equation is used to describe abrasive and adhesive wear but is not used for erosion, fretting, contact fatigue, or corrosive wear. Each type of wear has a different mechanism, cause, and effect.

8.6.2 Impact and Tribological Recommendations

The industry's focus has been on finding cost-efficient uses of LCA to inform decision-making. Decreasing costs and efforts come at the expense of the scope of the resulting streamlined LCA. Streamlined LCA requires a full understanding of the risks involved and trade-offs associated with streamlining. In the main, many of these academic versions of LCA are trying to facilitate the planned economies, which can be understood as economies that do not rely on market-based outcomes or solutions. The ongoing academic research has implications for LCA as a decision-support tool.

8.7 Brief Summary

In this chapter, the definition, disciplinary features, objectives, mission, technological connotations, and the future developing directions of green tribology have been comprehensively discussed.

9 Nanotribology

9.1 Introduction

Nanotribology is a branch of tribology, which involves the interactions between two relatively moving materials in contact at a nanometer or atomic scale, where the atomic forces are one of the most dominating forces. In macrotribology, tests are conducted on components with relatively large masses under heavily loaded conditions. Wear is inevitable and the bulk properties of mating components dominate the tribological performance. In micro-and nanotribology, measurements are made on components, with at least one having a relatively small mass, under lightly loaded conditions. Micro/nanotribological studies are essential for developing fundamental understanding of the interfacial phenomena on a small scale and studying interfacial phenomena involving ultrathin films and micro/nanostructures. These structures are commonly used in magnetic storage systems, micro-or nanoelectromechanical systems (MEMS/NEMS), and other industrial applications. The components used in micro-and nanostructures are very light and operate under very light loads.

Nanotribological and nanomechanics studies are crucial for developing a fundamental understanding of the interfacial phenomena on a small scale and studying interfacial phenomena in micro/nanostructures used in magnetic storage systems, MEMS/NEMS, and other applications. Friction and wear of the lightly loaded micro/nanocomponents are highly dependent on surface interactions (few atomic layers). Nanotribological studies are also valuable in the fundamental understanding of interfacial phenomena in macrostructures to provide a bridge between science and engineering.

9.2 Nano Friction

Studies of friction at an anoscopic scale have received great attention in recent decades. The development of MEMS/NEMS, for instance, requires a good understanding of the interfacial phenomena that significantly affect the performance of micro and nano-devices. Meanwhile, the invention of new scientific instruments, such as the Scanning Tunnel Microscope (STM), Atomic/Friction Force Microscope (FFM/AFM), Surface Force Apparatus (SFA), Quartz Crystal Microbalance (QCM), etc., and the rapid progress of computer simulation technology allow scientists to explore and resolve the secrets of adhesion and friction in more efficient ways than ever before. The main developments in nanotribology are show in Fig. 9 - 1.

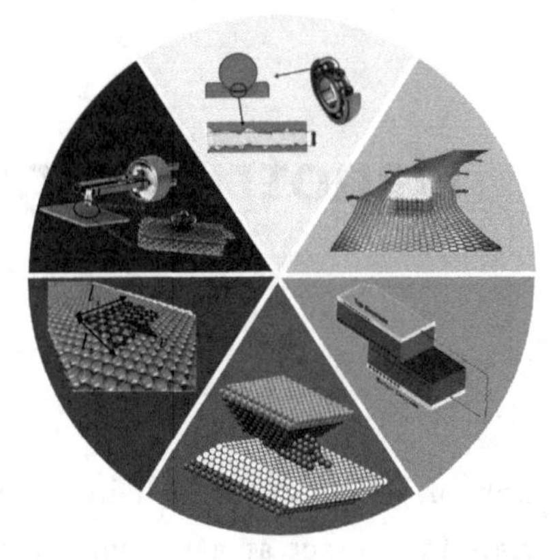

Fig. 9 - 1 Developments in nanotribology

9.2.1 Single atom friction and adhesion

To study friction mechanisms on an atomic scale, a freshly cleaved highly oriented pyrolytic graphite (HOPG) has been studied. The friction force maps after two-dimensional (2-D) spectrum filtering with high-frequency noise are truncated. The actual shape of the friction profile depends on the spatial location of the axis of tip motion. Note that a portion of the atomic-scale lateral force is conservative. The average friction force increases linearly with normal load and is reversible.

Everyday experience teaches us that adhesion and friction are related. Indeed, CoF between hard surfaces is classically cast as the sum of two terms, and one is due to the adhesion force and the other is due to the mechanics of plowing. The adhesion force can be obtained by measuring the force-distance curve or the 'approach-retraction curve'. The approach curve is the plot of the vertical cantilever bending versus the displacement of the rear end of the cantilever base. During retraction of the AFM probe, at point a, the probe snaps out of contact with the surface(Fig. 9-2). At this point, the tensile load

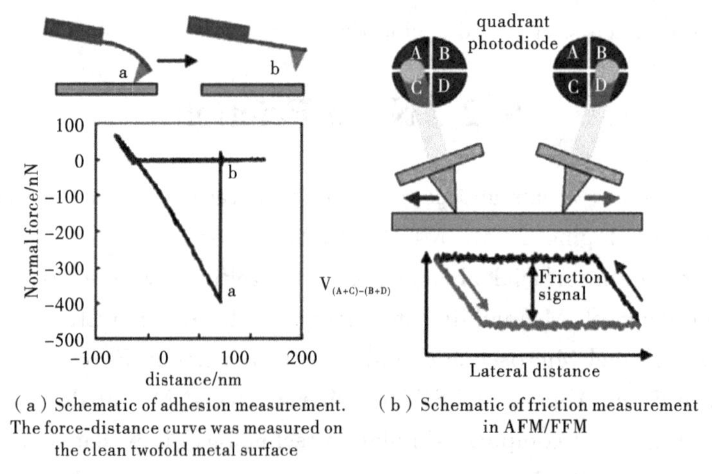

(a) Schematic of adhesion measurement. The force-distance curve was measured on the clean twofold metal surface

(b) Schematic of friction measurement in AFM/FFM

Fig. 9 - 2 The measurement process schematic of the force-distance curve of the adhesion force

equals the adhesion force of the tip-sample junction. Thus, the difference in force between a and b (free position) is attributed to the adhesion force. AFM is a good probe of adhesion force. In contacts between clean metals, however, plastic deformation generally occurs due to strong adhesion.

9.2.2 Surface Forces and Stick-Slip

During scanning, the tip moves discontinuously over the sample surface and jumps with discrete steps from one potential minimum (well) to the next. This leads to a sawtooth-like pattern for the lateral motion (force) with periodicity of the lattice constant. This motion is called as the stick-slip movement of the tip. The observed friction force includes two components-conservative and periodic, and nonconservative and constant. If the relative motion of the sample and tip are simply that of two rigid collections of atoms, the effective force would be a conservative force oscillating about zero. Slow reversible elastic deformation would also contribute to conservative force. The origin of the nonconservative direction-dependent force component could be phonon generation, viscous dissipation, or plastic deformation.

Some forces occur in vacuum, for example, attractive Van der Waals and repulsive hard-core interactions. Other types of forces can arise only when the interacting surfaces are separated by another condensed phase, which is usually a liquid. In vacuum, the two main long-range interactions are the attractive Van der Waals and electrostatic (Coulomb) forces. At small surface separations, the additional attractive interactions can be found such as covalent or metallic bonding forces. These attractive forces are stabilized by the hard-core repulsion. Adhesion forces are often strong enough to elastically or plastically deform bodies or particles when they come into contact. In vapors (e.g., atmospheric air containing water and organic molecules), solid surfaces in or close to contact will generally have a surface layer of chemisorbed or physisorbed molecules or a capillary condensed liquid bridge between them. Such a surface layer usually causes adhesion to decrease, but in the case of capillary condensation, the additional Laplace pressure or attractive capillary force may make the adhesion between surfaces stronger than in an inert gas or vacuum.

When totally immersed in a liquid, the force between particles or surfaces is completely modified from that in vacuum or air (vapor). The Van der Waals attraction is generally reduced, but other forces can now arise that can qualitatively change both the range and even the sign of the interaction. The attractive force in such a system can be either stronger or weaker than in the absence of the intervening liquid.

9.2.3 Effect of Lubricant Materials on Friction

Liquids are common lubricants and have been studied extensively at the macroscale. At the nanoscale, the tribological response of spherical liquid molecules has been well-characterized experimentally using surface force apparatus (SFA) and computationally with molecular dynamics (MD) simulations. When no external forces are applied to the system, the sliding stops and the solid-lubricant interactions are strong enough to force the liquid molecules and form a close-packed, ordered structure. The transformation of the liquid into this solid-like structure causes the two surfaces to effectively bond to each other through the lubricant. When the surfaces start to slide again, the lateral shear forces are introduced, steadily increasing, causing the molecules in the liquid to undergo small lateral displacements that change the film thickness. If these shear forces become greater than a critical value, the film disorders in a manner that is analogous to melting. In the case of sliding on insulating crystal surfaces, the solid-state lubricant may be in a superlubric state where the friction is negligible. In addition, when sliding occurs on metallic surfaces above the cryogenic temperatures, the electronic contributions to friction are no longer zero, and no superlubric state is possible. High applied pressures can force the fluid molecules out from between the two confining surfaces.

Nanoparticles are being considered for a wide variety of applications, including as fillers for nanocomposite materials, novel catalysts or catalytic supports, and components for nanometer-scale electronic devices.

9.2.4 The Odd-Even Effect

The link between friction and disorder in monolayers composed of n-alkane chains is examined using MD simulations. The tribological behavior of monolayers of 14 carbon atom-containing alkane chains, or pure monolayers, is compared to monolayers that randomly combine equal amounts of 12 and 16 carbon-atom chains or mixed monolayers. Pure monolayers consistently show lower friction than the mixed monolayers when sliding under repulsive (positive) loads in the direction of chain tilt. These MD simulations reproduce the trends observed in AFM experiments of the mixed-length alkanethiols and spiroalkanedithiols on Au. Harrison and coworkers have also examined the odd-even effect noted in experiments, where friction is found to be large for SAMs differing by one methylene group. The MD simulations demonstrate that the effect is due to conformational differences in the chains of different lengths, becoming more pronounced at higher loads.

9.2.5 Relation between Mechanical Properties and Friction

According to Bowden-Tabor's model, CoF (μ) can be expressed as $\mu = \mu_a + \mu_p$, where μ_a is the adhesion friction coefficient and μ_p is the plowing friction coefficient. In terms of the adhesion force, the mutual solid solubility of two contacting bodies can evaluate their tendency of adhesion. A higher mutual solid solubility of two contacting bodies causes a larger adhesion force, leading to a higher adhesion friction coefficient. If the adhesion between the film material and the tip is small, the change in the average

CoF with the annealing temperature and normal load is primarily determined by μ_p. Notably, μ_p is positively correlated with the penetration depth. This is because a higher penetration depth results in a bigger contact area between the film material and the submerged part of the tip, leading to a stronger obstacle against sliding. Meanwhile, the penetration depth is related to the hardness.

9.3 Nano Wear

If the normal force applied exceeds a critical value, which depends on the tip shape and on the material under investigation, the surface topography is permanently modified. In some cases, wear is exploited to create patterns with well defined shapes.

9.3.1 Atomic Scale Wear Phenomena

Atomically resolved images of the damage produced by scratching the FFM tip area on potassium bromide are obtained. A small mound that has piled up at the end of a groove on KBr (100) is shown at different magnifications. The groove is created a few minutes before imaging by repeatedly scanning with the normal force $F_N = 21$ nN; The image shows a lateral force map acquired with a load of≈ 1 nN, no atomic features are observed in the corresponding topographic signal. The debris extracted from the groove is recrystallized with the same atomic arrangement of the undamaged surface, which suggests that the wear process occurs in a similar way to epitaxial growth, assisted by the microscope tip. Although it is not easy to understand how wear is initiated and how the tip transports the debris, the important indications are given by the profile of the lateral force recorded while scratching. The mean lateral force multiplied by the scanned length gives the total energy dissipated in the process.

9.3.2 Nano Scratch

AFM can be used to investigate how surface materials can be moved or removed on micro to nanoscales, for example, in scratching and wear and nanofabrication/nanomachining. As expected, the scratch depth increases linearly with loads. Such microscratching measurements can be used to study failure mechanisms on the microscale and evaluate the mechanical integrity (scratch resistance) of ultrathin films at low loads.

Scratching can be performed under ramped loading to determine the scratch resistance of materials and coatings. CoF is measured during scratching, and the load at which CoF increases rapidly is known as the critical load, which serves as a measure of scratch resistance. In addition, post-scratch imaging can be performed in situ with the AFM in tapping mode to study failure mechanisms.

9.3.3 Nano Wear and Wear Mapping

A wear map is a chart that describes how a material's wear properties change under different stresses and strains. It contains several parameters such as CoF, wear rate, abrasive grain size, and so on. Creating a wear map requires a large number of experiments. By analyzing and processing the experimental data, the corresponding parameter values are derived and plotted on the chart. Analyzing the wear diagram provides a more comprehensive understanding of the wear characteristics of the material, which is significantly important for guiding the design and the selection of materials.

9.3.4 In situ Characterization of Local Deformation

In situ characterization of local deformation of materials can be carried out by performing tensile, bending, or compression experiments inside an AFM and by observing nanoscale changes during the deformation experiment. In these experiments, small deformation stages are used to deform the samples inside an AFM. In tensile testing of the polymeric films carried out, a tensile stage is used. The stage, equipped with a left-right combination lead screw (which helps to move the slider in the opposite direction) is used to stretch the sample to minimize the movement of the scanning area, which is kept close to the center of the tensile specimen. One end of the sample is mounted on the slider via a force sensor to monitor the tensile load. The samples are stretched for various strains using a stepper motor, and images are captured at different strain levels in the control area.

9.3.5 Nanoscale Indentation

For nanoindentation hardness measurements, the scan size is set to zero, and then a normal load is applied to make the indents using the diamond tip. During this procedure, the tip is continuously pressed against the sample surface for about 2 s under various indentation loads. The sample surface is scanned before and after scratching, wear, or indentation to obtain the initial and final surface topography, using a low normal load of 0.3 mN with the same diamond tip. An area larger than the indentation region is scanned to observe the indentation marks. Nanohardness is calculated by dividing the indentation load by the projected residual area of the indents. Direct imaging of the indent allows one to quantify the piling up of the ductile material around the indenter. However, it becomes difficult to identify the boundary of the indentation mark with great accuracy, making the direct measurement of the contact area somewhat inaccurate. A technique with dual capability of depth sensing as well as in situ imaging, which is most appropriate in nanomechanical property studies, is used for the accurate measurement of hardness with shallow depths. This nano/picoindentation system is used to make load-displacement measurements and subsequently carry out in situ imaging of the indent, if required. The indentation system consists of a three-plate transducer with the electrostatic actuation hardware used for the direct application of a normal load and a capacitive sensor used for the measurement of vertical displacement. The AFM head is replaced with this transducer assembly while the specimen is mounted on the PZT (piezoelectric cerarmic transducer) scanner, which remains stationary during indentation experiments. The transducer consists of a three-plate (Be-Cu) capacitive structure, with the tip mounted on the center plate. The upper and lower plates serve as drive electrodes, and the load is applied by providing an appropriate voltage to the drive electrodes. Vertical displacement of the tip (indentation depth) is measured by detecting the displacement of the center plate relative to the two outer electrodes using a capacitance technique. The indent area, and consequently the hardness value, can be obtained from the load-displacement data. The Young's modulus of elasticity is obtained from the slope of the unloading curve.

As the normal load increases, the indents become clearer, and the indentation depth increases. For the case of the hardness measurements at shallow depths on the same order as variations in surface roughness, it is desirable to subtract the original (unindented) map from the indent map for accurate measurement of indentation size and depth. To make accurate measurements of hardness at shallow depths, a depth-sensing nano/picoindentation system is used. Loading/unloading curves often exhibit sharp discontinuities, particularly at high loads. Discontinuities, also referred to as pop-ins, occurring during the initial loading part of the curve, mark a sharp transition from pure elastic loading to plastic deformation of the specimen surface, thus corresponding to an initial yield point. The sharp discontinuities in the unloading part of the curves are believed to result from the formation of lateral cracks at the base of the median crack, causing the surface of the specimen being thrusted upward. The hardness of single-crystal silicon and single-crystal aluminum at shallow depths on the order of a few nm (i.e., on the nanoscale) are found to be higher than at depths on the order of a few hundred nanometers (i.e., on the microscale). Microhardness has also been reported to be higher than that at the millimeter scale by several investigators. The data reported show that hardness exhibits scale (size) effects.

9.4 Nano Lubrication

9.4.1 Effect of Chain Length of Lubricants

The influence of the lubricant carbon chain length on its performance is significant. Straight-chain alkanes can improve the viscosity and friction reduction performance of lubricants, but their antioxidant and anti-pollution properties are poor. Branched alkanes can enhance the antioxidant, corrosion resistance, and anti-wear properties of lubricants, but they also tend to increase cost. Cycloalkanes and aromatic hydrocarbons can improve the antioxidant and anti-pollution properties of lubricants, while also improve their anti-wear properties and stability. For different uses and working conditions, it is very important to develop appropriate lubricant formulations.

9.4.2 Solid Superlubricity

Solid superlubricity mainly comes from the synergistic effect generated by the relative sliding of weakly interacting surfaces in incommensurate contact, and the interfacial atomic synergism can effectively reduce the sliding barrier. It should be pointed out that the incommensurability interfaces do not necessarily produce superlubricity with ultra-low friction, and the magnitude of friction is closely related to the degree of incommensurability of the structure, the rigidity of the structure, and the interactions. Although the stacking the incommensurate interfaces can effectively reduce the sliding energy barrier, it also naturally causes the contact between surfaces to be in an unstable state with high potential energy. This unstable structure is prone to the structural distortion and deformation under the effect of thermal activation and pollutant adsorption, etc., and returns to the local intimate contact state, leading to the destabilization or even loss of superlubricity behavior.

9.4.3 Liquid Superlubricity

Liquid lubrication is one of the main methods of reducing friction. Unlike the friction mechanism described in solid superlubricity, the main source of friction in liquid lubrication is the internal friction of the fluid, which, according to lubrication theory, can be reduced by lowering the viscosity of the lubricant. However, as the viscosity of the lubricant decreases, so does its load-carrying capacity, making it impractical to significantly reduce friction in this way. Therefore, tribologists have been working in this field for a long time, trying to find the practical ways and effectively reduce friction.

9.5 Tribochemistry in Nanotribology

Tribochemistry is a subset of mechanochemistry. Although both of these terms are well known and their specific reaction effects, especially mechanochemical ones have been widely discussed for over a century, even in the past decade, it is stated that the precise nature of these reactions is still not well understood. The most significant achievement is the differentiation of the effect of mechanical stress from that of heat, which leads to the establishment of a new branch of chemistry called 'mechanochemistry', where materials react differently from thermal reactions. Tribochemistry is a branch of chemistry that deals with the chemical and physico-chemical changes of solids due to the influence of mechanical energy.

9.5.1 Velocity and Temperature Dependence

To investigate the effect of velocity on friction and adhesion, the friction force versus normal load at different velocities is measured. The variation of the friction force, adhesive force, and CoF as a function of velocity is summarized, the friction force decreases logarithmically with increasing velocity. Velocity has very little effect on the friction force, which reduces slightly only at very high velocity. The mechanisms of the effect of velocity on adhesion and friction can be explained. At high velocity, the meniscus is broken and does not have enough time to rebuild. The contact stresses and high velocity lead to tribochemical reactions of the Si(100) wafer and Si_3N_4 tip, which have native oxide (SiO_2) layers, with water molecules. $Si(OH)_4$ is removed and continuously replenished during sliding. The $Si(OH)_4$ layer between the tip and the Si (100) surface is known to have low shear strength and causes a decrease in the friction force and CoF in the lateral direction. The chemical bonds of Si-OH between the tip and the Si(100) surface induce a large adhesive force in the normal direction.

9.5.2 Surface Roughness

To obtain roughness-independent friction, lateral or torsional modulation techniques are used, in which the tip is oscillated in-plane with small amplitude at a constant normal load, and the change in the shape and magnitude of the cantilever resonance is used as a measure of the friction force. These techniques also allow the measurements over a very small region (a few nm to a few mm). As mentioned earlier, the shift in contact resonant frequency in both the lateral and TR modes is a measure of contact stiffness. At an excitation voltage above a certain value, as a result of microslip at the interface, a flattening of the resonant frequency spectrum occurs. At low excitation voltage, the AFM tip sticks to the sample surface and follows the motion like an elastic contact with the viscous damping, in which case the resonance curve is Lorentzian with a well-defined maximum. The excitation voltage should be high enough to initiate a microslip. The maximum torsional amplitude at a given resonance frequency is a function of the friction force and sample stiffness, so the technique is not valid for inhomogeneous samples. If the torsional stiffness of the cantilever is very high compared with the sample stiffness, the technique should work.

9.5.3 Environment

To study the effect of the environment, friction tests of the friction pair are conducted in high vacuum, argon, dry air (less than 2% RH), air with 30% RH, and air with 70% RH. By comparing the CoF in different environments, it is found that the friction in argon is the lowest for the SAMs tested. In a high vacuum, the intimate contact leads to high friction. In dry air, the friction is higher than in argon. This shows that oxygen has an apparent effect on the performance of SAMs. Kim et al. studied the thermal stability of alkylsiloxane SAMs in air. It is found that the alkylsiloxane decomposes at about 200 ℃, which is much lower than the decomposition temperature of 470 ℃ in a vacuum. This difference could be attributed to the oxygen in the air. The water contained in the air is found to have a significant influence on the friction of SAMs. A study on the effects of humidity on alkylsilane on a mica substrate indicates that water molecules can penetrate the alkylsilane film, which alters their molecular chain ordering and can also detach alkylsilane molecules from the substrate. A summary of the coefficients of friction before the failure of the lubricating films in various environments is presented. The data are average values based on five measurements. To summarize the highlights, friction of the tested lubricant films is high in high vacuum because of the intimate contact between the lubricants and the counterpart surface. Friction of the tested lubricant films is lower in argon than in dry air. Friction of the SAMs is significantly influenced by water molecules.

9.6 Atomic-Scale Computer Simulations

Analytic models and computational simulations have played an important role in characterizing and understanding friction. Most analytic models divide the complex motions that create friction into more fundamental components defined by quantities such as spring constants, the curvature and magnitude of potential wells, and bulk phonon frequencies. In atomic-scale molecular dynamics (MD) simulations, the atom trajectories are calculated by numerically integrating coupled classical equations of motion. Interatomic forces that enter these equations are typically calculated either from total energy methods that include the electronic degrees of freedom, or from simplified mathematical expressions that give the potential energy as a function of interatomic displacements. MD simulations can be considered numerical experiments that provide a link between analytic models and experiments. The main strength of MD simulations is that they can reveal unanticipated phenomena or unexpected mechanisms for well-known observations.

9.6.1 Molecular Dynamics Methodology

Molecular dynamics simulations are straightforward to describe: giving a set of the initial conditions and a way of mathematically modeling interatomic forces, Newton's (or equivalent) classical equation of motion is numerically integrated. The forces acting on any given atom are calculated, and then the atoms move a short increment forward in time in response to these applied forces. This is accompanied by a change in atomic positions, velocities, and accelerations. The process is then repeated for some specified number of time steps. The output of these simulations includes new atomic positions, velocities, and forces that allow additional quantities such as the temperature and pressure to be determined. As the size of the system increases, it is useful to render the atomic positions in the animated movies that reveal the responses of the system in a quantitative manner. Quantitative data can be obtained by analyzing the numerical output directly.

There are several different approaches by which interatomic energies and forces are determined in MD simulations. The most theoretically rigorous methods are those that are classified as ab initio or first principles. These techniques, which include density functional theory and quantum chemical ab initio methods, are derived from quantum mechanical principles and are generally both the most accurate and the most computationally intensive. They are, therefore, limited to a small number of atoms (<500), which has limited their use in the study of friction. Alternatively, the empirical methods are functions containing parameters that are determined by fitting experimental data or the results of ab initio calculations. These techniques can usually be relied on to correctly describe 10 empirical methods that have, therefore, been widely used in studies of friction. Semi-empirical methods, including tight-binding methods, include some elements of both empirical methods and ab initio methods. For instance, they require the quantum mechanical information in the form of, for example, on-site and hopping matrix elements, and include fits to experimental data. Empirical methods simplify the modeling of materials by treating the atoms as spheres that interact with each other via repulsive and attractive terms that can be either pairwise additive or many-body in nature. The repulsive and attractive functional forms generally depend on interatomic distances and/or angles and contain adjustable parameters that are fit to ab initio results and/or experimental data.

9.6.2 Friction at Atomic Scale

Work is required to slide two surfaces against one another. When the work of sliding is converted to a less ordered form, as required by the first law of thermodynamics, friction will occur. For instance, if the two surfaces are strongly adhering to one another, the work of sliding can be converted to damage that extends beyond the surfaces and into the bulk. If the adhesive force between the two surfaces is weak, the conversion of work results in damage that is limited to the area at or near the surface and produces transfer films or wear debris. While the thermodynamic principles of the conversion of work to heat are well known, the mechanisms by which this takes place at sliding surfaces are much less well established despite their obvious importance for a wide variety of technological applications. Atomic-scale simulations of friction are therefore important tools for achieving this understanding. They have consequently been applied to numerous materials in a wide variety of structures and configurations, including atomically flat and atomically rough diamond surfaces, rigid substrates covered with monolayers of alkane chains, perfluorocarboxylic acid and hydrocarboxylic

Langmuir-Blodgett (LB) monolayers, between contacting the copper surfaces, between a silicon tip and a silicon substrate, and between contacting diamond surfaces that have the organic molecules absorbed on them.

Sliding friction that takes place between two surfaces in the absence of lubricant is termed dry friction, even if the process occurs in an ambient environment. Simple models have been developed to model dry sliding friction that, for example, considers the motion of a single atom over a monoatomic chain. Results from these models reveal how elastic deformation of the substrate from the sliding atom affects energy dissipation and how the average frictional force varies with changes in the force constant of the substrate in the direction normal to the scan direction. Much of the correct behavior involved in dry sliding friction is captured by these types of simple models. However, more detailed models and simulations, such as MD simulations, are required to provide information about more complex phenomena.

MD simulations have been used to study the sliding of metal tips across clean metal surfaces by numerous groups. An illustrative case is shown for a copper tip sliding across a copper surface. Adhesion and wear occur when the attractive force between the atoms on the tip and the atoms at the surface becomes greater than the attractive forces within the tip itself. Atomic-scale stick and slip can occur through nucleation and subsequent motion of dislocations, and wear can occur if part of the tip gets left behind on the surface. The simulations can further provide data on how the characteristic stick-slip friction motion can depend on the area of contact, the rate of sliding, and the sliding direction.

An additional study of stick-slip in the sliding of much larger, square-shaped metal tips across metal surfaces is carried out using EAM potentials. This study predict that the collective elastic deformation of the surface layers in response to sliding is the main cause of the stick-slip behavior shown. The simulations also predicts that the stick-slip produces phonons that propagate through the surface slab.

9.6.3 Elastic Properites at Nanoscale

The nature of adhesive interactions between a clean, deformable metal tip indenting metal surfaces has been identified and clarified over the last decade through the use of MD simulations. In particular, the high surface energies associated with the clean metal surfaces can lead to strongly attractive interactions between surfaces in contact. This wetting mechanism was first discovered in MD simulations and has been confirmed experimentally using the AFM.

9.7 Application and Development of Nanotribology

Nanotribology research has a wide range of application needs. With the development of the precision machinery and high-tech equipment, especially the emerging disciplines driven by nanotechnology, such as nanoelectronics, nanobiology, and the study of microelectromechanical systems involves peripheral friction and surface interface behaviors. Due to scale and surface effects, the laws that these problems follow are no longer based on macroscopic tribological principles.

9.7.1 Nanotribology of MEMS Devices

Small and lightweight micro-electro-mechanical systems (MEMS) and nanoelectro-mechanical systems (NEMS) are gaining increasing importance in aerospace applications, integrated sensors, and intelligent control systems. In these systems of highly integrated electronics, mechanical components are co-fabricated on planar wafers and subsequently etched free for the mechanical movements in three dimensions. Si is a well-established material for the fabrication of MEMS and NEMS but the reliability of Si moving components has been in doubt as Si is a brittle material in nature. MEMS and NEMS usually have a large surface area-to-volume ratio that makes them particularly vulnerable to surface damage due to high friction and adhesion of the components to the adjacent structures during operation or use. A major design limitation for these systems is their inability to withstand prolonged sliding surface contact due to high friction. The short functional lifetime of these devices is attributed to the high CoF and excessive wear rate of the devices. Due to the small dimensions of MEMS and NEMS, wear debris accumulated at the interface can cause jamming. Wear can also lower the corrosion resistance of the components in a harsh environment. It is also noticed that CoF in a micro or nanodevice is greater than that in a macro device due to the size effect under similar operation conditions. Therefore, friction is a more critical issue for NEMS and MEMS. Another important issue associated with MEMS and NEMS is adhesion that arises during the physical contact of the components when a MEMS or NEMS device is in operation. To overcome those problems, many researchers have used surface modification techniques or dry/liquid lubrication methods.

9.7.2 Nanotribology of Magnetic Storage

Magnetic storage devices used for storage and retrieval are tape, flexible (floppy) disk, and rigid disk drives. These devices are used for audio, video, and data storage applications. In the data storage industry, magnetic rigid disk drives media, tape drives/media, flexible disk drives/media, and optical disk drive/media are used. Magnetic recording and playback involve the relative motion between a magnetic medium (tape or disk) against a read-write magnetic head. Heads are designed so that they develop a (load-carrying) hydrodynamic air film under the steady operating conditions to minimize head - medium contact. However, physical contact between the medium and head occurs during starts and stops, referred to as contact-start-stops technology. In modern magnetic storage devices, the flying heights (head-to-medium separation) are on the order of 5-20 nm, and roughnesses of head and medium surfaces are on the order of 1-2 nm RMS. High stiction (static friction) and wear are the limiting technologies for the future of this industry. High stiction and wear are the major impediments to the commercialization of the contact recording. The most commonly used thin magnetic films for tapes are evaporated Co-Ni (82-18 at.%) or Co-O dual layer. Typical magnetic films for rigid disks are metal films of cobalt-based alloys (such as sputtered Co-Pt-Ni, Co-Ni, Co-Pt-Cr, Co-Cr, and Co-NiCr). For high recording densities, trends have been to use thin-film media.

9.8 Brief Summary

Nanotribology is a branch of tribology, which involves the interactions between two relatively moving materials in contact at a nanometer or an atomic scale. Tribology research exceedingly needs the broadened knowledge in various fields such as physics, chemistry, mechanics, materials science, etc.

10 Tribology Under Extreme Environments

With the development of science and technology, the load, operating speed, temperature, and other working conditions of various friction pairs have become increasingly stringent. The study of tribological problems under extreme working conditions has become a subject with the social and economic benefits.

10.1 Introduction

Friction leads to a large amount of mechanical energy loss, and the research on anti-friction and anti-wear technology is becoming more and more in-depth. Wear failure is closely related to the tribological materials. The properties of friction pair materials and lubricants largely determine the extreme operating conditions and wear characteristics of the tribological system. Conventional lubrication technology and tribological materials are far from meeting the requirements of modern high-tech industries such as aerospace and aviation for the use of the tribological systems operating under extreme environments. Lubricating materials such as solid lubricating materials and high-temperature lubricating oils/greases are a new type of lubricating materials that have emerged under this background. It effectively breaks through the usage limits of lubricating materials under conventional working conditions, making it possible to design tribology for extreme working conditions such as high temperature, high vacuum, high load, high speed, strong radiation, and special working media. At the same time, it also provides more material selection for the prevention of wear failure, and lays the research foundation for the prevention of modern mechanical wear failure. This chapter mainly introduces the tribological behavior characteristics and research methods under various extreme environments, and tries to be of benefit to engineers and technicians engaged in the tribological design and wear failure prevention under extreme working conditions, as shown in Fig. 10 - 1.

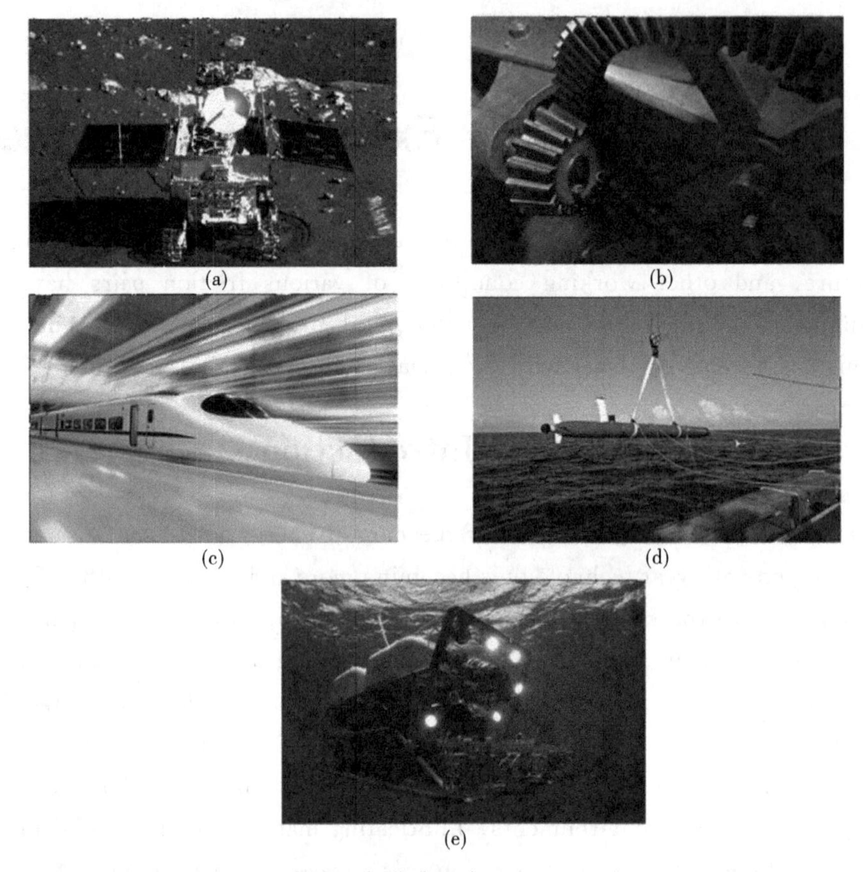

Fig. 10 - 1 Cases of tribology in extreme environments

10.2 High Vacume Tribology

Vacuum is an important consideration that affects the friction and wear properties of materials in space applications. In tribological components, the presence of these adsorbed layers prevents the intimate metal-to-metal contact thereby reducing adhesion and friction. In the vacuum of space, any physisorbed layers are quickly removed by evaporation and oxide layers are worn away.

10.2.1 Solid Lubrication

Solid lubricants are low-shear-strength materials whose easy-shear properties confer low friction when presents at sliding interfaces. Graphite is the most well-known solid lubricant and widely used in industry, but in a vacuum, it is worn rapidly as its lubricity is sustained only in the presence of water vapor or certain other condensable vapors. As such, it is unsuitable for space applications, and other low-shear materials such as the metal dichalcogenides, soft metals, and polymers, are preferred. Solid lubrication is broadly implemented by two routes: ①thin lubricant coatings, ②self-lubricating materials (such as certain types of cages in ball bearings).

1. TMD

Transition metal sulfide (TMD) materials, such as WS_2, MoS_2, and NbS_2, are a type of solid lubricant with a hexagonal lattice structure. The shear strength in the direction parallel to the crystal plane is very low, and it is easy to slide, so it has good tribological properties in a vacuum environments. MoS_2 is a solid lubricant used in a vacuum environment, mainly suitable for gears, rolling bearings, and cams, etc. The wear life in vacuum is dozens of times longer than that in air, and the CoF in a vacuum is about one-fourth of that in air.

WS_2/MoS_2 multilayer film has better friction performance than MoS_2 single-layer film in a vacuum. In the case of ambient air, the CoF of the single-layer film and the multi layer film is about 0.1, while in the case of vacuum, the CoF is less than half of this value. In particular, CoF of the multilayer film in vacuum is close to 0.03, and the life is more than twice that of ambient air. In vacuum, almost no influence of humidity or oxygen is observed, and both single-layer and multilayer films exhibit lower CoF and higher friction durability than those in ambient air.

2. DLC films

Diamond-like Carbon (DLC) film is a metastable amorphous carbon film. The carbon atom orbitals contain in it combining with sp^2 and sp^3 hybrids. Therefore, it combines some of the excellent properties of diamond and graphite, such as high hardness, good chemical stability, and biological compatibility, excellent infrared light transmission performance, ultra-low CoF, and excellent wear resistance. The CoF of DLC coating in a vacuum (6.5×10^{-3} Pa) is lower than in ambient air (CoF 0.08). At the same time, the ball pairing shows that a C-rich transfer film is formed, which is the main reason why DLC has a good lubricating effect in a vacuum environment. Vacuum cuts off oxygen, so DLC can exert good lubrication performance in a vacuum environment.

3. Metals and metal oxides

Titanium alloy has low density, high specific strength, excellent corrosion resistance, and good heat resistance. It has become an important structural material in aerospace, chemical energy, and other industrial fields. However, the low load-bearing capacity and poor tribological properties limit the application range of titanium alloys. The poor tribological properties of titanium alloys are closely related to their low hardness, low plastic shear resistance, and poor surface oxide protection. Taking Ti_6Al_4V as an example, the dry friction and wear mechanism in air is mainly controlled by plastic deformation and oxidation. At low sliding speeds, oxidative wear and micro-cutting are the main wear mechanisms. At the same time, titanium oxide can also be used as a lubricant under vacuum. Pure titanium alloy undergoes an anodic oxidation process to form a porous oxide layer on the surface. Under ambient air and vacuum conditions (pressures of 10^{-3} mbar and 10^{-6} mbar), the hard anodized coating obtains the best wear resistance.

Aluminum and aluminum alloys have a series of excellent physical, chemical, and mechanical properties, such as low density, high plasticity, good electrical conductivity, low price, and easy recycling. However, aluminum alloys also have shortcomings such as softness, easy adhesion, high CoF, poor wear resistance, and difficult lubrication, which limits their wide applications to a certain extent. Alumina has good wear resistance due to its high hardness, but poor lubrication performance. Tests carried out under a 2×10^{-5} Pa vacuum clearly show that the softening of the surface layer caused by radiation damage leads to a significant reduction of CoF by approximately two times.

Ferroalloy has low production costs and large-scale output, and it is the most widely used metal material in the industry. The contradictions and disputes on the friction and wear of iron-based alloys mainly focus on the formation mechanism of slight wear and severe wear. CoF in air is 0.8, and CoF in vacuum drops to about 0.6. CoF is lowest at 0.05 MPa. There are oxidative wear characteristics; the steel disk is worn more severely in a vacuum, and the steel ball hardly suffers from the wear, but there is a small amount of material transfer.

4. Polymer

The polymer organic coating is mainly used as an electrothermal barrier layer, a thermal protection layer, a conductive layer, shock absorption and sound insulation layer, a waterproof and moisture-resistant layer, a temperature-resistant and corrosion-resistant layer, a radiation-resistant layer, and a layer with excellent mechanical properties. PTFE coating has good lubricity, low CoF, and other characteristics, and has a wide range of application prospects in the field of the space technology. Environmental pressure has an insignificant effect on the CoF of PTFE coating. CoF of PTFE coating under air conditions is slightly higher than that under vacuum conditions. PTFE has the advantage that it can produce low friction in both air and vacuum but because of its low tensile strength and it has to be reinforced with glass fiber. Without reinforcement, it is unlikely to withstand the stresses induced by the combination of the normal applied load and the tangential frictional force and will wear rapidly. PTFE is commonly used as a cage material in the form of a PTFE/glass fiber/MoS_2 composite in otherwise unlubricated ball bearings or those coated with sputtered MoS_2. These polyimide composites are an alternative bearing cage material.

10.2.2 Fluid Lubrication

The use of oils in spacecraft mechanisms is in general limited to those that exhibit exceedingly low volatility such asperfluoropolyethers (PFPEs), multiply alkylated cyclopentanes (MACs), and polyalphaolefins (PAOs). Such oils are evaporated at remarkably low rates at(and below)room temperature. However, at high temperatures, the depletion of oil by evaporation may be such as to prevent it from surviving sufficiently long to meet a mechanism's operational life.

PFPE oil has an exceptionally low vapor pressure and offers a lower rate of evaporative loss than oil. In PFPE-lubricated steel contacts, fluorine is released from the oil by the action of repeated shearing under high contact stresses and reacts with iron in the steel to form iron fluoride, FeF_3.

PFPE grease has good chemical inertness and very low vacuum volatility, so it is widely used in the lubrication of space-moving mechanism parts such as bearings, harmonic reducers, ball screws, etc. Greases offer the following advantages over oils: ①They can act as sealants to prevent or minimize the ingress of foreign matter into bearings. ②They can trap and retain wear particles. ③Some grease thickeners can act as boundary lubricants. ④Some grease thickeners can act as sponges for storing the oil and delivering it to ball-race contacts (note that PTFE filler does not act as a sponge).

Ionic liquid refers to a salt composed of positive and negative ions that are liquid at or near room temperature and also called room temperature ionic liquid, room temperature molten salt, etc. There are no electrically neutral molecules in the ionic liquid, and 100% is composed of anions and cations. It can be classified according to the difference of anions and cations. Ionic liquids are non-flammable and explosive, with low melting points, low volatility, good oxidation resistance, and high thermal stability. These characteristics are very consistent with the expected performances of an ideal lubricant, which makes it has the potential to become a lubricant for space machinery and is expected to become a high-performance lubricant for special lubrication under harsh conditions such as aviation and computer industries. The ionic liquid exhibits excellent friction-reduction, and antiwear proprieties, both in air and vacuum.

10.2.3 Tribo-Components and Their Lubrication

The need for longer spacecraft missions and the requirement for mechanisms to undergo more complex motion profiles have pushed existing space lubricants to their limits. The integrity of tribological components and the manner of their lubrication are keys to successful space mechanism design. Bearings and other tribological components in space applications have requirements additional to those of terrestrial mechanisms. Designs must be sufficiently robust to survive launch loads and launch-induced vibrations, and they must allow the efficient mechanism operation over many years in the space environment. As the environment can entail thermal and vacuum extremes, great care is needed to ensure that the selected lubricants can be effective and remain so over the often lengthy lifetime of the spacecraft.

In practice, the choice of space lubricants is limited to a small number of low volatility oils (PFPEs, MACs, and PAOs) and a few types of vacuum compatible solid lubricants (MoS_2, lead, PTFE, and polyimides). All lubricants have a finite life, and their ability to meet ever-increasing mission lifespans is being pushed to the limit. Longer life requirements will hasten the development of lubricant replenishment systems. The miniaturization of the mechanical systems brings further lubrication challenges, while nano-scale devices offer the prospect of operation in a super-low-friction regime.

10.2.4 Vacuum and Lubrication

At present, there are three main types of space lubrication, namely solid lubrication, semi-solid lubrication, and liquid lubrication.

Solid lubrication refers to the use of solid powders, films, or some integral materials to reduce the friction and wear between two bearing surfaces. In the lubrication process, solid lubricants and the surrounding medium physically and chemically are reacted with the friction surface and formed a solid lubricating film, thereby reducing friction and wear. Under high vacuum conditions in space, mineral oil is volatile, which not only causes poor lubrication, but also can easily cause contamination. Therefore, in the early stage of the development of space lubricants, solid lubricants are primarily used, which are also one of the main lubrication methods for spacecraft and space stations at this stage.

Grease lubrication is often used under normal pressure and vacuum conditions. It is often used in the lubrication of bearings under low vacuum and air-sealed structures. It is easy to use and has a large viscosity at low temperatures and a long life.

Liquid lubrication has a long service life, and the number of cycles can reach more than 1×10^9. It is mostly used for high vacuum, medium, and high-speed bearing lubrication, and especially suitable for precision and sensitive rolling bearings. Liquid lubrication entails small starting torque, but its lubrication system is complicated, and the lubricating oil replenishment method and seal design need to be considered.

10.3 Tribology at Extreme Temperatures

Aviation products generally undergo rigorous testing under high and low temperature environments (−55 - 125 ℃). The space environment usually includes ultra-high vacuum, extreme temperature, and temperature alternation, atomic oxygen, particle radiation, microgravity, etc. When flying on the sunny side of space, the aircraft has to withstand strong radiation from the sun. The surface temperature can reach 200 ℃ or higher. When flying on the shaded side, the surface temperature can be as low as −200 ℃. This requires the spacecraft to be able to operate normally in extreme temperatures and large temperature differences.

10.3.1 Tribology at Elevated Temperature

Oils and greases can reduce friction between sliding contacts primarily by providing velocity accommodation between surfaces in relative motion by shearing of the oil molecules across the solid-liquid-solid interface. However, most liquid lubricants are volatilized at high temperatures, which may lead to the failure of the lubricated component and potentially affect other unrelated components when the vapor from the lubricant condenses or reacts on their surfaces. Solid lubricants are utilized for high-temperature applications since their vapor pressure is low, and hence, sublimation does not contribute to the degradation of the system components. CoF is around 0.2 in most applications. In addition to CoF, the wear rate is also an important factor in the design of the moving assemblies. Wear on sliding surfaces is typically the result of one or more of the following main mechanisms: ①abrasive wear. ②adhesive wear. ③fatigue wear. ④chemical wear. Typical wear rate values for solid lubricants with moderate wear resistance range between 10^{-6} and 10^{-5} $mm^3 \cdot (N \cdot m)^{-1}$. Solid lubricants fall into one of two categories: ① intrinsic lubricants (e. g., MoS_2 and soft metals). ② extrinsic lubricants (e. g., graphite, which needs to be terminated with water or some other condensable vapor). Solid lubricants currently used at high temperatures may be divided into three categories: ①soft metals (e.g. Ag, Cu, Au, Pb, and In). ②fluorides(e.g., CaF_2, BaF_2, and CeF_3). ③metal oxides (e.g. V_2O_5, $Ag_2Mo_2O_7$). All the three types of materials undergo plastic deformation and/or form low-shear-strength surfaces at elevated temperatures.

Two examples of applications of particular interests to high-temperature friction reduction are: ① air foil bearings for high mach aerospace engines. ② high-speed machine tools.

Airfoil bearings in high mach engines are designed to operate under high loads in air at 900 ℃ well-suited for oxide lubrication. They rely on an self-generated, hydrodynamic air layer that forms above a critical rotation velocity, as the bearing changes shape with the formation of an interfacial boundary layer. The quest for high-performance solid lubricants over a broad temperature range has been strongly influenced by the challenging operating conditions of airfoil bearings.

High speed machining allows for rapid production of finished parts. Self-hardening mechanisms in TiAlN coatings can occur over extended periods at high temperature owing to spinodal decomposition, further increasing the wear resistance. The properties of these thin-film compounds are sometimes modified further with one or more elements, such as vanadium, known to form Magnéli phases at very high temperatures (>600 ℃), demonstrating additional properties of thermal stability, low adhesion, and easy shear.

High-temperature lubricants are made of synthetic base oils and high-performance additives to produce ashless synthetic ester high-temperature chain lubricants, which do not contain substances that can form solid residues. They are used for lubricating bearings, chains, slide rails, and gears where the ambient temperature often reaches 260 ℃. In order to further improve the tribological properties of the alloy matrix at high temperatures, researchers use a variety of solid lubricants to reduce the CoF, such as DLC and TMD materials.

The high-temperature friction behavior of alloys is affected and controlled by the self-generated oxide film on the surfaces. Sliding contact on rough surfaces is a complex process, accompanied by the generation of friction heat, thermo-elastoplastic deformation, and thermo-elastic instability, representing a typical thermo-mechanical coupling problem.

Generally, high-temperature wear is divided into high-temperature sliding wear, erosion, fretting, and abrasive wear. The adhesion mechanism emphasizes the thermal softening of the metal caused by the increase in temperature. The essence of adhesion is the formation of metal bonding on the surface when the heated softened metal contacts, and local metal transfer and macroscopic particle shedding occur on the wear surface. The abrasive wear mechanism posits that oxides act as abrasive particles causing abrasive wear. The increase in temperature leads to surface oxidation exacerbating the wear process.

10.3.2 Tribology in Cryogenic Environment

Until now, interest in low-temperature technology has been restricted to applications in space technology and superconductivity. One example is the utilization of liquid hydrogen as an environmentally friendly energy carrier for transportation and energy supply systems.

Many working conditions in space technology and the aviation industry exceed the usage limit of lubricating oil and grease, which puts forward high requirements for solid lubrication technology. The friction system of some mechanical parts outside the space station, the drive of the solar cell array, the expansion and contraction mechanism, the high-power brush-slider, etc., must ensure reliable lubrication and longevity. Rocket engines, space shuttles, heavy loads in ultra-low temperature and strong active environments, lubrication of high-speed bearings, and sealing from ultra-low temperature to high temperature under high pressure are all key technologies.

Friction systems (such as bearings, seals, valves, etc.) in low-temperature environments often generate excess heat and severe wear. Traditional lubricating oils or lubricating esters lose their function at low temperatures. Because the low-temperature range is far below the pour point of the base oil, only solid lubricants or materials with good tribological properties can be used.

Low temperature has a great influence on the crystal structure and organization of the material surface. The effect of the temperature decrease on the wear life of solid lubricating coatings is unclear and is determined by the changing physical and mechanical properties of the solid lubricating coating under cooling, as well as the rate of the adhesive friction failure process. At low temperatures, components with interacting surfaces in relative motion, such as bearings or valves, and tribosystems in general, are widely used. Because of the strongly temperature-dependent properties, conventional lubricants, such as oils and greases, are not applicable. Additionally, the environment in low temperature systems affects the interactions of the elements. For example, lubrication systems working by tribochemical reactions with the environmental humidity cannot be used because any water is frozen out. In inert gases like helium or nitrogen, the environmental conditions are similar to those of a vacuum. In hydrogen or oxygen, however, the materials can react with the environment, like with the consequence of changes in material properties. Coatings are widely used for the wear protection or friction reduction at room temperature and above. The material properties of polymers are strongly temperature-dependent. Young's modulus and hardness are much higher at low temperatures compared to room temperature, whereas the already low heat conductivity continues to decrease. While at ambient temperature adhesive wear dominates, abrasive wear is observed at low temperatures.

10.3.3 Design of Lubricants under Wide Temperature

Early research found that a single solid lubricant has a high temperature lubrication and anti-friction effect, but it always operates within a certain temperature range. Peterson et al. investigated the high-temperature tribological properties of a large number of oxides and found that except for a few oxides such as PbO, which have lubricity in a wide temperature range, other oxides are used as lubricants at a very narrow temperature range. Studies have shown that some common solid lubricants (such as MoS_2 and graphite, etc.) are easily oxidized and failed at high temperatures. Although other solid lubricants have good oxidation resistance and low CoF at high temperature, their CoF at low temperatures is very high. Therefore, research on high-temperature materials, especially those with good anti-friction and wear resistance in a wide temperature range, has become a research hotspot in the field of materials and solid lubrication.

10.4 High Speed Tribology

With the rapid increase in the operation speed of friction pairs, the contact surface temperature caused by the high speed greatly increases, leading to a decrease in the friction and wear performance of the contact surface, and a series of deterioration phenomena such as thermal fatigue, thermal cracking, thermal oxidation, and thermal adhesion of the material occur. The degree of this deterioration and the critical conditions are related to many factors such as the service conditions of the friction pair, the structure of the material, and the thermal performance of the material.

10.4.1 High-Speed Friction Applications

Since tribology was formally proposed in the 1960s, it has attracted much attention and has become one of the fastest-growing disciplines. With the development of the mechanical equipment and transportation toward high speeds and the increased safety, many friction pairs are operated under special working conditions such as high speed and high temperature, leading to friction characteristics that differ from those under general working conditions. In aviation, chemicals, machinery, military, and other industries, the relative sliding speed of the friction surface often exceeds 40 m/s. For example, the sliding speed of the traveler in the ring spinning machine on the ring runway can reach 45 m/s or high. The sliding speed in some weapon launch systems can be as high as hundreds of meters per second or higher. For this reason, tribology researchers at home and abroad have launched research on the high-speed friction properties of materials.

10.4.2 Friction Material at High Speed

1. Metal materials

At present, iron-based materials are the most widely used materials in high-speed equipment. Tribological properties of the carbon and nitriding layer on the surface of carbon steel are studied under high-speed dry friction conditions.

2. Fiber reinforced ceramic matrix composite

Fiber-reinforced ceramic matrix composite material is a general term for materials composed of fiber as reinforcement, and the fiber-reinforced ceramic matrix is combined through a certain composite process. This type of composite material has high strength, high toughness, excellent thermal and chemical stability, and is a new type of structural material. If the fiber-reinforced ceramic composite material is under the action of solid lubricating substances (such as graphite and sulfide), the self-lubricating property of the lubricating substance can reduce the friction factor or reduce the fragmentation of the ceramic composite material.

3. Ceramic material

With the continuous improvement of transportation speed, more and more stringent requirements have been put forward on the performance of key components of braking technology, such as brake disc (brake disc hub material, brake shoe, brake disc) and brake pad. The essence of brake tribology is sliding dry friction under high speed and high contact pressure. Ceramic-reinforced aluminum matrix composites have the advantages of lightweight, wear resistance, and good thermal conductivity, and their application potential in the field of high-speed vehicle braking has attracted widespread attention.

10.4.3 Experimental Equipment, Research Methods

The conventional friction and wear test is carried out on a friction testing machine. In the experiment, the friction torque, time, speed curve, dynamic friction factor, wear rate, etc. are measured, and then the sample is micro-analyzed. At the same time, there are different experimental techniques for materials with different application conditions, such as intermittent tests and continuous tests for brake materials, as well as mathematical simulation methods.

10.5 Tribology in the Ocean Environment

The ocean is the carrier on which mankind depends for survival, and it shares with the future development of mankind. Friction and wear in extreme ocean environments is one of the key issues restricting the application and promotion of ocean materials.

The deep ocean is a harsh environment, especially with regard to its high hydrostatic pressure (approximately 1 MPa per 100 m depth). The harsh environment may induce friction, wear, and strength degradation in different materials, which greatly influence the performance and life of deep-ocean equipment. One of the key requirements of deep-ocean equipment is that the selected engineering material should exhibit excellent friction and wear resistance under high hydrostatic pressure.

10.5.1 Characteristics of Ocean Environment

The ocean environment refers to the broad expanse of the continuous seas and oceans on Earth and its unique natural environment, including seawater, substances dissolved and suspended in seawater, seabed sediments, ocean organisms, and ocean atmospheric environment. The ocean environment is complex, leading to the complex and diverse tribological problems faced by ocean equipment.

The viscosity of seawater is very low, about 1/20 to 1/100 that of mineral oil, and the lubrication performance is very poor. It is difficult to form an effective elastohydrodynamic lubricating film on the surface of a friction pair. Seawater is a typical corrosive electrolyte, which contains not only a large amount of chloride, but also carbonate that is often saturated, and a large amount of magnesium and calcium ions. The electrical conductivity is hundreds of millions or even tens of billions times higher than mineral oil, which can cause electrochemical corrosion of most metal materials and chemical aging of most polymer materials. The ocean environment is extremely complex, often in the conditions of wind and waves, heavy rains, and ocean changes. It is easy to produce turbulence, vibration, and impact on ocean equipment, causing abnormal wear on some of the connectors and key friction pairs. The essence of corrosion, friction, and wear in ocean equipment is the reduction of contact materials, deterioration of performance or failure, which leads to the reduction or failure of the reliability of marine structures or parts. The tribological challenge of materials in the ocean environment involves studying the corrosion behavior, microbial adhesion corrosion behavior, and electrochemical corrosion behavior of materials in the harsh marine environment based on the characteristics of the ocean environment and the resultant tribological problems.

10.5.2 Tribology of Metal Materials

1. Corrosive Wear

Seawater is a typical corrosive electrolyte. The metal structures or parts exposed to the ocean environment undergo electrochemical reactions with the surrounding medium and are severely corroded. In seawater, abrasive wear, plastic deformation, and corrosive wear are dominant. Seawater plays a role in cooling, lubricating, and causing corrosion to the friction pairs. However, under large loads, the seawater environment will significantly increase the wear rate. The wear mechanism is mainly the combined effect of micro-plowing, plastic deformation, and corrosion.

2. Erosion

Erosion is usually caused by fluid scouring, leading to the localized acidification of the material, preventing the formation of the protective films or causing thinning and cracking of the material surface's passivation film or protective film. This process may also induce plastic deformation of the material, local energy increases, thereby accelerating the corrosion of the material.

3. Alternating Load

The cyclical movement of water in the ocean, such as wind, waves, currents, and tides, greatly accelerates the friction and wear of key components, such as tidal and wind power generation devices, oil and gas drilling tools, and ship power systems. Studying the friction and wear mechanism under the combined action of corrosion and alternating loads will provide a better understanding of the tribological behavior of friction pairs in the oceanic environment.

10.5.3 Tribology of Polymer Materials

Compared with metal materials, polymer materials have excellent corrosion resistance, self-lubricating properties, and the ability to embed abrasive or sand particles, and often have good tribological properties. They have a very wide range of applications in marine equipment, such as ship stern bearings, sealing devices for marine equipment, water-lubricated bearings, and pistons.

The tribological behavior of polyetheretherketone (PEEK), polyphenylparaben (PHBA), polyimide (PI), and perfluoroethylene propylene copolymer (FEP) sliding with GCr15 and 316 steel rings under seawater lubrication has been studied and compared with pure water lubrication. The results show that the friction and wear behavior of the polymer under the lubrication of the aqueous medium is not only related to the performance of the polymer itself, but also related to the corrosive and the lubricating effects of the medium.

10.5.4 Tribology of Ceramic Materials

Ceramic materials are inorganic non-metallic materials, which have the advantages of high hardness, high temperature resistance, corrosion resistance, high rigidity, minimal thermal expansion coefficient, good thermal conductivity, high strength, wear resistance, and environmental friendliness, and have broad application prospects. With the continuous promotion and application of new engineering ceramic materials in the engineering field and their applications in the harsher special working environment, seawater lubricated ceramic bearings are developed under this background.

(WAl) C-Co/fluoride (CaF_2, BaF_2, CaF_2/BaF_2) self-lubricating ceramic composites are prepared by mechanical alloying and hot pressing sintering, and their tribological behavior under seawater environment is studied. It can be found that fluoride composite has good tribological properties in seawater. The CoF of the composite material under a load of 60 N is 0.15, with a corresponding wear rate of 1.3×10^{-5} $mm^3/(N\cdot m)$.

10.5.5 Lubrication and Anti-Wear

At present, many marine tribological devices and components are made of metal alloys, which are prone to corrosion and wear. Therefore, it is of great significance to develop new friction materials with excellent anti-wear and anti-corrosion properties as a substitute for marine metal tribological materials.

As a corrosion-resistant, self-lubricating polymer material, PEEK has great prospects in the tribological research in ocean environments. The addition of carbon fiber (CF) can greatly improve the wear resistance of PEEK under seawater lubrication. In particular, the 10 vol% CF reinforced PEEK composite has the highest wear resistance, even under high load conditions.

10.6 Multi-Field Coupling Tribology

All tribological phenomena happening near interfaces between solids are determined by the atomic interactions within and between solids, as well as those between atoms of the substances present at the interface. Since these interactions give rise to various physical effects described at the macro scale by different theories and models, the tribological interface can be considered a 'paradise' of Multiphysics. The following types of phenomena may take place in such an interface or in its immediate vicinity: mechanical (solid and fluid), thermal, electro-magnetic, metallurgical, quantum, and others.

10.6.1 Tribology in Radiation Environment

With the rapid development of aerospace technology, required service life of spacecraft is increasing, and reliability is becoming more and more important. Space environment is still the main cause of various abnormal and even emergency failures of different spacecraft. In the low Earth orbit (LEO, 200-1000 km altitude) cosmic environment, atomic oxygen (AO) is the main environmental species, which is mainly a negative factor. Past research and space mission reports indicate that many materials undergo oxidation and corrosion due to AO radiation and cosmic environment exposure. Therefore, more attention should be paid to the response behavior of spacecraft materials in the AO environment.

10.6.2 Tribology in Space Temperature Environment

Space environment usually includes ultra-high vacuum, extreme temperature, and temperature alternation, atomic oxygen, particle radiation, microgravity, etc. For the internal moving parts of space machinery, the influence of temperature and vacuum is great. Due to the influence of the orbit, attitude, azimuth, structure, and surface properties, the spacecraft is in a complex environment of vacuum and high temperature during operation.

10.6.3 Tribology in Complex Media Environment

The thermal environment can vary greatly, from cryogenic temperatures (for deep space missions and those involving cooled infrared devices) to the elevated temperatures (250 ℃ for Mercury-orbiting spacecraft). Few lubricants remain effective when exposed to this combination of high vacuum and temperature extremes. Radiation, comprising intense fluxes of electrons and protons, is another environmental factor that can degrade lubricants, however, lubricants are shielded from potentially harmful effects by the surrounding metal. Research has demonstrated that the wear rate is closely related to temperature. Additionally, studies have shown that in high-speed sliding with large current, oxidation of carbon material directly affects wear due to the high temperature.

Although many systems involve two or more of these phenomena at various scales, multiscale and multiphysics models are still challenging to develop and use, as they require multidisciplinary expertise and collaborative efforts. Breakthroughs are thus expected from the future development of versatile and efficient multiscale/physics tools dedicated to tribology.

10.7 Brief Summary

This chapter introduces the tribological behaviors of several typical materials under some extreme conditions.

11 Tribology of Basic Components

The tribological practices are extremely important from the view point of reliable and maintenance free operation of machine components.

11.1 Introduction

Tribology encompasses how interacting surfaces and other tribol-elements behave in relative motion in natural and artificial systems. There is mutual movement and interaction between and within the parts of the machine in the work, so there are tribological problems between the parts.

The bearings are divided into sliding bearings and rolling bearings.

11.2 Journal and Thrust Bearing

11.2.1 Journal Bearing

The plain bearing is simplest type of bearing, comprising just a bearing surface with no rolling elements. The simplest example of a plain bearing is a shaft rotating in a hole. A simple linear bearing can be a pair of flat surfaces designed to allow motion; for example, a drawer and the slides, it rests on or the ways on the bed of a lathe.

Plain bearings are divided into radial plain bearings and thrust plain bearings according to the loading direction (Fig. 11 - 1). Radial plain bearings mainly support

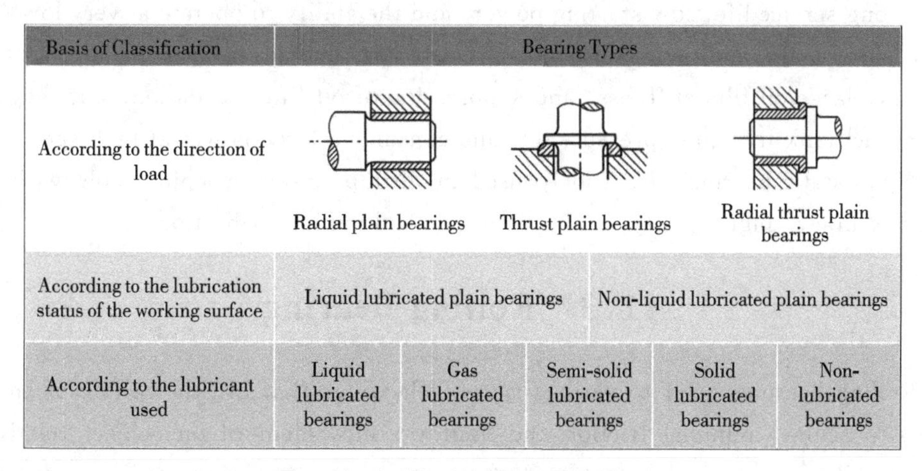

Basis of Classification	Bearing Types				
According to the direction of load	Radial plain bearings		Thrust plain bearings		Radial thrust plain bearings
According to the lubrication status of the working surface	Liquid lubricated plain bearings			Non-liquid lubricated plain bearings	
According to the lubricant used	Liquid lubricated bearings	Gas lubricated bearings	Semi-solid lubricated bearings	Solid lubricated bearings	Non-lubricated bearings

Fig. 11 - 1 Types of journal bearings

radial loads. Thrust plain bearings can only support the axial loads. Friction performance between sliding surfaces of plain bearings varies along with different working conditions and lubrications. It can generally be divided into the complete fluid friction, boundary friction, and dry friction. The fluid may be oil, water, gas, or other media. Damage forms of sliding bearings are as follows: scratch, abrasive wear, adhesive wear, fatigue wear, peel, corrosion wear, and fretting wear.

11.2.2 Thrust Bearing

Thrust bearings are generally composed of two thrust gaskets or more and a number of rolling bodies. Generally, thrust gaskets are divided into shaft plates and seat plates. The most common type of rolling bodies is generally made of iron or copper cages. The most common type of bearing is steel ball thrust bearing. Thrust bearings are designed to support thrust loads and their surfaces are perpendicular to the axis of rotation, They contain different pads to ensure the satisfactory lubrication.

11.2.3 Squeeze Film Bearing

Oil film bearing is a kind of precision bearing. In fact, it is a hole with very small inner surface roughness. When the shaft rotates at high speed, a thin layer of lubricating oil (because the lubricating oil has a certain tension) supports the suspended shaft. When the shaft is at rest, it makes contact with the bearing. For example, the oil bearing in an electric fan motor is a simple oil film bearing.

11.2.4 Hydrostatic Bearing

Hydrostatic bearings refer to sliding bearing that rely on an external supply of the pressurized oil to create a hydrostatic bearing oil film inside the bearing, achieving liquid lubrication. Hydrostatic bearings work under liquid lubrication, resulting in minimal wear, long service life, low starting power, and the ability to operate at very low (even zero) speeds. In addition, this bearing also has the advantages of high rotation accuracy, large oil film stiffness, and suppression of oil film oscillation, but they need special fuel tank to supply pressure oil, and consume a large power at high speeds.

Hydrostatic bearings are widely used in ultra-precision machine tools with high rotary accuracy, high rigidity, stable rotation, and minimal vibration.

11.3 Rolling Bearing

Rolling bearings refer to various types of bearings that use the rolling of balls or rollers to achieve minimal friction and limit the movement of one object relative to another. They are generally composed of an inner ring, outer ring, rolling body, and a cage. According to the direction of the load they can bear, rolling bearings can be divided into the centripetal bearing, angular contact bearing, and thrust bearing. According to the types of rolling body, they can be divided into ball bearings and roller bearings(Fig. 11 - 2).

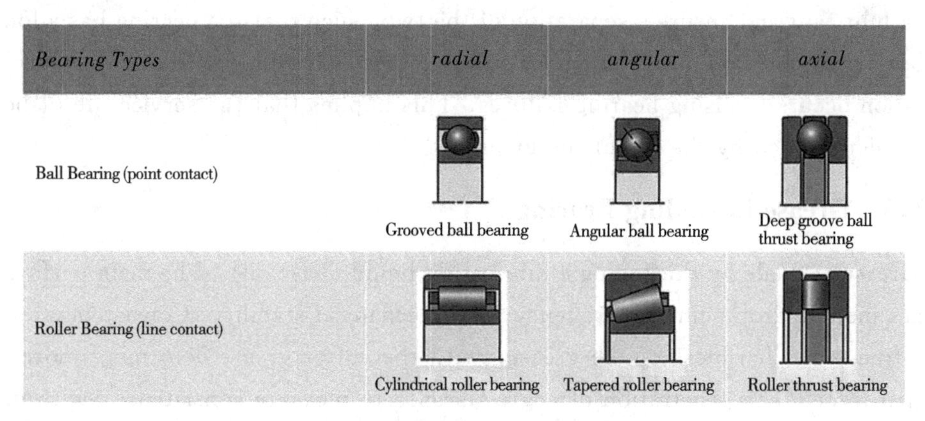

Fig. 11 - 2 Types of roller bearings

11.3.1 Rolling Bearing Materials

Steel for bearings includes steel for rolling bearing parts and other auxiliary materials. The inner and outer rings and rolling bodies of bearings are mainly made of high carbon chromium bearing steel. Common materials are carburized bearing steel, stainless bearing steel, high-temperature bearing steel, and medium-carbon bearing steel. In addition, for bearings used under special working conditions, the corresponding special performance requirements of the steel should be specified, such as high-temperature resistance, high-speed performance, corrosion resistance, and magnetic resistance.

11.3.2 Fluid Lubricants in Rolling Bearing

The fluid lubricants with a certain viscosity are used to separate the surfaces from each other and avoid wear. There are many methods of lubrication. All methods serve three necessary goals: ① providing a sufficient film to prevent or mitigate asperity interaction, ②removing or distributing locally generated heat, ③minimizing debris in the contact zone. The lubricants can be grease, oil, or a solid lubricant. Rolling bearings are commonly lubricated by oil lubrication and grease lubrication. Roughly 90% of all rolling bearings are grease lubricated. In some special environments (such as high temperature, vacuum), solid lubricants can also be used for lubrication.

For ball bearings lubricated with oils or greases, the degree of separation, and therefore wear, of the balls and races is dependent on speed. At low speed, there is little or no separation, and contact occurs between balls and race, this is known as the boundary regime. At very high speeds, the hydrodynamic lift ensures full separation of balls and races, and there is no ball-to-race contact whatsoever, this corresponds to the elastohydrodynamic (EHD) regime. At intermediate speeds, mixed lubrication can occur where metal-to-metal contact is intermittent.

Main role of grease in a rolling bearing is to provide the rolling element-ring contact with a lubricant and ensure a separation of the two, such that the bearing has a long life and low friction. Main disadvantage of using grease is its limited life. Severe lubricant starvation occurs, causing bearing failures. This implies that the service life of bearing may be determined by the life of the grease.

11.3.3 Grease in Rolling Bearing

Grease is made by thickening a lubricating liquid (base oil). The main performance indexes include drop point, consistency, and mechanical stability. Consistency refers to the softness and hardness of the grease, with the softer grease deforming more under external force. The penetration of degree is used to measure consistency, with greater penetration indicating softer grease and smaller consistency. Mechanical stability refers to the ability of grease to resist mechanical damage in use.

Grease lubrication has a simple structure, easy to seal, has high strength of oil film, resistance to lose, and has certain ability to prevent water, gas, dust, and other harmful impurities from invading bearings. Therefore, it is widely used in general circumstances. But the viscosity of grease is large and they usually generate significant heat at high speeds, making it more suitable for low-speed applications.

11.3.4 Failure and Wear of Rolling Bearing

Wear is the process of continuous damage of surface materials in the relative motion of objects in contact with each other. It is the inevitable result of friction. The wear classification method expresses people's understanding of wear mechanisms, and different scholars have put forward different classification views.

In 1953, Sovieut classified wear into the following three categories according to the effect of friction surfaces: mechanical wear, molecular-mechanical wear, and corrosion-mechanical wear.

Wear in rolling bearings is a non-linear phenomenon, and even observations and measurement of rolling bearing wear in regular intervals show that the simple intuition is not sufficient to predict how wear will evolve over time. This is due to a variety of reasons: wear depends mainly on local sliding and local load, and wear itself modifies.

11.4 Gear

Gear transmission is a very important mechanical transmission. The main characteristics of gear transmission are: it can transfer the motion and power between any two axes in the space, and the applicable power and circumferential speed range is wide.

Successful operation of gear depends on the provision of extreme pressure oils, which are oils containing additives that the formed surface-protective layers at elevated temperatures. Lubricants used for gear lubrication usually contain surface-active additives, and the prevailing mode of lubrication is mixed or boundary lubrication. Therefore, wear is typically mild and probably corrosive as a result of the action of boundary lubrication.

11.4.1 Wear and Failure Modes of Gear

Under normal circumstances, the failure of gear transmission is mainly attributed to the failure of gear teeth. Failure of gear transmission is related to the working conditions and service conditions. In terms of working conditions, gear transmission has two types: closed gear transmission and open gear transmission. Gear of closed transmission is enclosed in the box, which can ensure good lubrication. Important gear transmissions generally adopt closed transmission. Open drive gears are exposed to air, dust, sand, and other particles, making it easy for them to enter, leading to poor lubrication. They are usually used in manual, low-speed, and other unimportant gear transmission. Failure of gear also has many forms of gear tooth fracture, tooth surface pitting erosion, tooth surface bonding, tooth surface wear, tooth surface plastic flow and so on (Fig. 11 - 3).

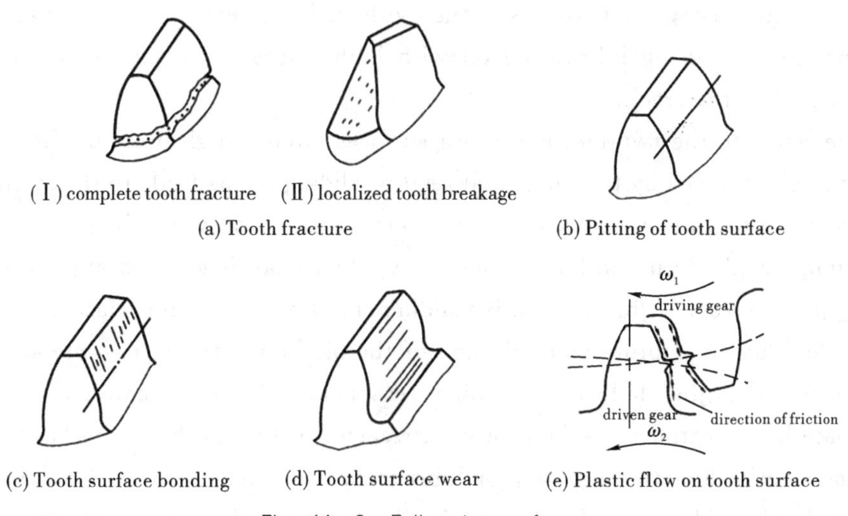

Fig. 11 - 3 Failure types of gear

Tooth fracture refers to the integral or partial fracture of one or more teeth of a gear. Fracture usually occurs at the root of the tooth. There are fatigue fractures and overload fractures. When the gear teeth engage, they begin to be stressed. Maximum bending stress and stress concentration will occur at the rounded corners of the teeth, just like the cantilever beam. Bending stress disappears after the tooth is removed. Therefore, when the gear is working, root of each tooth is subjected to varying bending stress. When bending stress exceeds fatigue limit of the material, the root of the tooth will produce fatigue cracks, cracks continue to expand, and finally lead to tooth fracture. This fracture is called fatigue fracture.

If the gear made of brittle materials such as cast iron and integral hardened steel is subjected to significant impact force or severe overload during operation, the gear teeth may break suddenly. This fracture is called overload fracture. To improve the fracture resistance of gear teeth, measurements can be taken as follows: increasing the radius of the tooth root fillet to reduce the influence of the stress concentration on fatigue strength. The shot peening and other strengthening techniques are employed at the tooth root. Utilizing proper materials and heat treatment to toughen the tooth core and harden the surface, to improve the anti-breaking ability of the tooth.

Pitting is a kind of failure of gear transmission resulting from the pitting damage of the tooth surface caused by the spalling of small pieces of metal. When gear teeth enter meshing, contact stress of the tooth surface is approximately fluctuating, where contact stress exceeds the contact fatigue limit stress of gear tooth material, fatigue cracks will occur under the surface of the gear tooth, which continuously expands over time and forms pitting corrosion. Pitting generally occurs on the surface near the nodal line near the root of the tooth. In open gear transmission, pitting is seldom found because the tooth surface is worn quickly. The pitting resistance of the tooth surface can be improved by increasing the hardness of the tooth surface, reducing the roughness of the tooth surface, and using lubricating oil with high viscosity to reduce the invasion of lubricating oil in the crack.

Glue refers to the two meshing tooth surfaces, under high pressure direct contact adhesion, accompanied by the tangential relative sliding. This leads to the tearing off of metal from the tooth surface, resulting in a serious adhesion wear phenomenon.

Gluing usually occurs in high-speed, heavy-duty gear drives. Local pressure at the meshing place is very high, and relative sliding speed of the tooth surface is low, which will also lead to the rupture and adhesion of the oil film between two metal surfaces, thus causing the glue failure. Matching reasonably the gear material, the use of appropriate heat treatment method for improvement of the tooth surface hardness, low speed and overloading the viscosity of lubricating oil, high speed and overload when mixed with glue additives of lubricating oil, reducing the tooth surface roughness implemented, the shift gears are adopted to decrease the addendum methods of relative sliding velocity, all can improve the ability of tooth surface of agglutination.

In open gear transmission, dust, sand, or other hard debris enters between the meshing tooth surfaces and causes wear on the tooth surfaces under rolling action of the teeth. When a hard tooth with a rough surface meshes with a softer tooth, the softer tooth surface also develops wear. After excessive wear and tear, teeth of the wheel will thin out and break when they reach a certain point.

The most effective way to reduce or prevent tooth wear is to use a closed transmission. In addition, it can also improve the hardness of the tooth surface, reduce the surface roughness, keep the lubricating oil clean, and the regular replacement to prevent or reduce the tooth surface wear.

For gear transmission with heavy loads and low speeds, if surface hardness of the gear tooth is low, the material of the gear surface may slip along the direction of friction under the action of great friction force, forming the phenomenon that the active gear tooth surface is raised near the pitch line, which is called the plastic flow of gear surface. To improve the hardness of the tooth surface, lubricating oil with greater viscosity is chosen, and the plastic flow of the tooth surface is reduced and prevented.

11.4.2 Gear Lubrication

1. Lubrication status

Good lubrication can make the tooth surface maintain a complete layer of lubricating oil film. The state of tooth surface lubricating film is closely related to tooth surface contact stress, sliding direction, and lubricating oil performance.

(1) Boundary lubrication state: Most gear transmission devices of metallurgical equipment are in a boundary lubrication state.

(2) Extreme pressure lubrication: Extreme pressure additives are added to the oil to improve lubrication performance. It can react with the tooth surface to produce extreme pressure lubrication film. Therefore, good extreme pressure lubricating oil must be used to maintain good extreme pressure lubrication.

(3) State of solid lubrication film: This lubrication state is the solid lubricant powder or coating agent with the tooth surface, which keeps the tooth surface with a layer of solid lubrication film. Molybdenum disulfide and graphite are widely used as solid lubricants.

2. Lubrication mode

(1) Oil bath lubrication: Oil bath lubrication with gearbox as an oil tank, the gear immersion in a certain depth of oil, due to the rotation of gear, stirring oil splash, oil droplets splash to each part of the lubrication. Oil bath lubricated cylindrical gear drive, the circumferential speed is generally not more than 12 – 15 m/s. The circumferential speed of the worm gear and worm drive generally does not exceed 6 – 10 m/s.

(2) Circulating lubrication. Cycle lubrication is a separate lubrication system. The oil is pumped through pump to the gearbox, then back to the mailbox, and so on. It is suitable for gear transmission with high circumferential speed and high power. Circular lubrication is required for cylindrical gear with circular speed greater than 12 – 15 m/s and worm with circular speed greater than 6 – 10 m/s.

(3) Grease lubrication. Some gears with low speed and heavy loads are lubricated with grease, which is proven to be effective in reducing wear and avoiding oil leakage. At present, in metallurgical equipment, some gearboxes are lubricated with semi-fluid

grease.

(4) Solid lubrication. The cylindrical gear reducer load is not large, and relatively smooth, and the semi-dry film of molybdenum disulfide for lubrication is used. This lubrication method has a good effect.

3. Selection of lubricants for gear transmission

There are usually several principles to choose from:

(1) Gear load is the main basis for selecting oil products. Light loads can choose oil without additives. For high loads with significant sliding, such as worms, can choose oil containing oil additives. For heavy load surface and strong impact, such as hyperbolic gear, rolling mill gear seat, the use of full extreme pressure lubricating oil should be considered.

(2) Gear speed is the main basis for selecting oil viscosity. Select low-viscosity oil with high speed, and choose high viscosity oil with low speed.

(3) Lubrication mode is the reference condition of oil selection. Circulating lubrication must require good fluidity of oil products, and cylinder oil containing more viscous should not be selected. Cylinder oil can be used for oil bath lubrication.

11.4.3 Gear Life

Gear fatigue performance, especially contact fatigue resistance, is directly and prominently impacted by the conditions of the interacting gear surfaces, such as lubrication and surface topographies, since the distributions of the contact pressure and the resultant stresses/strains can be significantly affected. Gear fatigue life and reliability relate closely to lubrication properties. The gear surface life increases with the rise of the lambda ratio, namely the improving lubrication condition. Superfinishing can significantly enhance the gear surface fatigue resistance, resulting in a long surface fatigue life.

Anti-wear and extreme-pressure oil additives and high viscosity lubricants provide great benefits for gear surface fatigue life. Oil degradation may induce failures in the form of white etching or dark etching regions. Gear lubricants with appropriate high viscosity can reduce vibration and noise. The lubricant temperature rise that decreases oil viscosity can cause rattle oscillations.

11.5 Mechanical Seal

Mechanical seal refers to a device that prevents fluid leakage by keeping close and sliding relative to at least one pair of end faces perpendicular to the rotating axis, under the action of fluid pressure, elastic force (or magnetic force) of the compensation mechanism, and auxiliary seal(Fig. 11 - 4).

In light seals, rubber bellows are also used as auxiliary seals. Rubber bellows have limited elasticity and generally need to be supplemented by springs to meet the loading elasticity. A mechanical seal is a device that helps join systems or mechanisms together by preventing leakage (e.g., in a plumbing system), containing pressure, or excluding contamination. A stationary seal may also be referred to as 'packing' modern cartridge seal designs do not damage the pump shaft or sleeve. Hydraulic seals are generally non-metallic and largely made up of materials like thermoplastic elastomers and rubbers. The hydraulic seal works in static as well as in dynamic conditions. The basic work is to provide the proper sealing and prevent contaminants from entering the system.

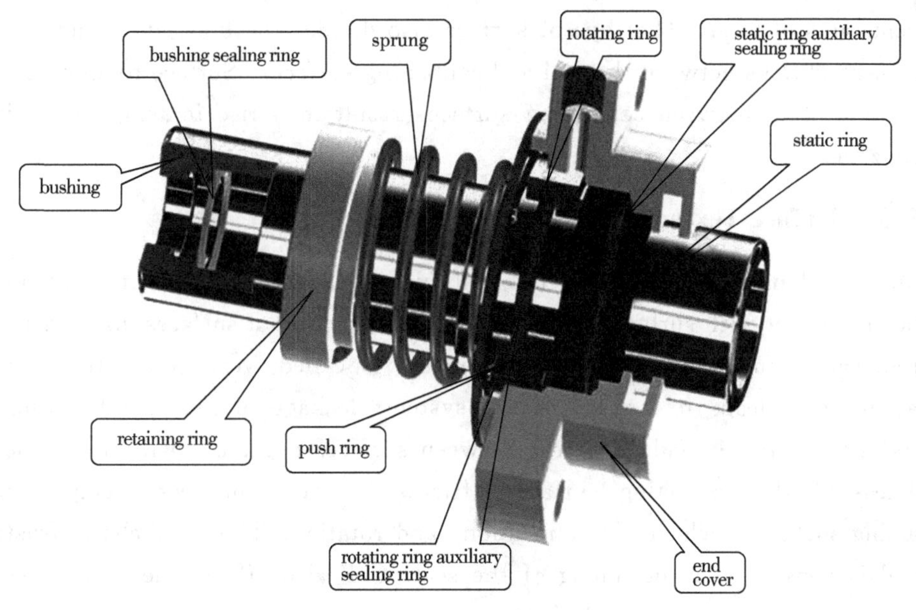

Fig. 11-4 Meachanical seal

11.5.1 Theory of Seal

Mechanical seal functions by maintaining close contact and sliding relative to the shafts end face perpendicular to the shaft axis, under the influence of fluid pressure and the elasticity a compensation mechanism of (or magnetic force), along with the aid of auxiliary seals, to achieve leakage resistance in shaft sealing.

A commonly used mechanical seal structure is composed of the static ring, rotating ring (moving ring), elastic element spring seat, set screw, rotating ring auxiliary sealing ring, static ring auxiliary sealing ring, and other components. An anti-swivel pin is fixed on the gland to prevent the rotation of the static ring. Rotating rings and stationary rings are also often referred to as compensating rings or non-compensating rings, depending on whether they have axial compensating capability.

Isotropic surfaces are rough, which can enhance wear on these surfaces. Dry contact occurs due to the interaction between surface asperities. Anisotropic surfaces tend to have lower friction and experience fluid-asperity interaction (mixed hydrodynamic lubrication conditions) between the seal surface and contacting surface. Continuous wear changes the geometry of the sealing surface, affecting both the CoF between the two surfaces and the contact pressure distribution between them. Consequently, leakage of the seal is affected, leading to seal failure.

11.5.2 Surface Treatment of Mechanical Seal

The wear of seal material also depends on the thermal expansion of the seal, frictional heat, and angular speed. Isotropic surface and anisotropic surface have different rates of wear. The type of surface also decides the lubrication and frictional behavior conditions between the seal and contacting surfaces. Surface roughness profile and high friction between contacting surfaces result in a rise in temperature in the sealing zone.

11.5.3 Surface Texture of Mechanical Seal

In most dynamic applications, the seal surface experiences both translational and rotational contact with surfaces in contact. The interaction of surfaces in contact causes the removal of material from the seal surface is called wear. Over the period of operation, wear leads to leakage of the system, leakage of pressure build-up, and results in a severe seal failure. Wear between surfaces depends on the properties and roughness of the contacting surfaces, friction between surfaces, temperature of contacting surfaces, velocity of translation, and rotation. The wear characteristics of the seal is considered in designing of the seal. CoF also affects the wear rate. CoF increases as the material removal rate increases and as aging time increases. In dynamic operation, initially, the CoF may be low but tends to increases as the operation proceeds.

11.6 MEMS

MEMS is classified as: ①sensor-based, which has sensing elements, ②actuator-based, which has elements that undergoing mechanical motion. In these devices, tribological issues such as adhesion, friction, and wear strongly manifest, which undermines the mechanical motion of MEMS elements. Conventionally, thin films/coatings have been researched for their application to MEMS as solutions to mitigate these tribological issues. Examples include self-assembled monolayers (SAMs), diamond-like carbon (DLC) coatings, polymers, and perfluoropolyether films (PFPE). In recent years, bioinspired approaches have attracted attention as better alternative solutions for MEMS.

11.6.1 Tribology in MEMS Fabrication

MEMS is built at the micro/nano-scale. At these scales, the ratio of surface area to volume is high. Hence, body forces such as inertia and gravity become insignificant. In contrast, surface forces such as capillary, Van der Waals, electrostatic, and chemical bonding dominate. These surface forces cause adhesion at the interface of contacting MEMS elements. Amongst these forces, the capillary force that arises due to the condensation of water from the environment is the strongest. Further, adhesion strongly influences friction at the micro/nano-scale. MEMS is traditionally made from silicon due to availability of the process knowledge developed for the material in semiconductor industries. However, silicon does not have good tribological properties. Silicon due to its inherent hydrophilic nature experiences high surface forces, and because of its brittle nature, it undergoes severe wear. Thus, the improvement of the tribological performance of silicon is the key to realizing the smooth operation of actuators-based MEMS.

11.6.2 Tribology in MEMS Operation

There are clearly tribological elements involved, as all devices involve turning joints at which friction and wear must be controlled and minimized. Wear is a particular concern in any arrangement. Again, tribology is key in both the operation and the possible failure modes of the device.

1. Surface engineering in MEMS—diamond like carbon films

DLC is a term used to describe a family of carbon based materials, many of which have attractive tribological properties—such as low CoF and low wear rates, high values of the elastic modulus, chemical inertness, and good stability. When discussing the tribological properties of DLC, as with anything else, it is important to take account of the nature of the counterface. For a number of different counterface materials (including steel, silicon nitride, and sapphire), the long-term stability appears to be dependent on the formation of a carbon-rich transfer layer derived from the DLC but with a distinctive morphology of its own, often consisting of fine graphitic nano-particles (<5 nm) within a distorted diamond-like structure.

2. Thin films

Engineers and scientists are aware that at small scales, forces such as capillary force, Van der Waals force, chemical bonding, and electrostatic contributes to adhesion, which in turn significantly influence friction. Among these, capillary force is the strongest, which forms a meniscus bridge as a result of environmental water condensation. In addition, MEMS devices prohibit the use of conventional liquid and solid lubrication. Therefore, the real challenge is to minimize the surface forces and the occurrence of wear in small scaled devices. Many tribologists across the globe have reported enhancing the silicon performance by adapting various surface modification techniques, including topography and chemical modifications. Topography modification includes the roughening surfaces and surface texturing. It may also involve imitating the lotus leaf surface topography (Lotus Effect) by modifying the local geometry of surfaces to obtain from getting wet by the encompassing water. With respect to the tribological enhancement through chemical modification, there is an incredible search for both natural and inorganic coatings and lubricants. While many coatings show low initial friction, and in contact wear tests, few exhibit high wear strength.

It is clear that tribology is an important factor affecting the performance and reliability of MEMS devices. MEMS materials need to exhibit good tribological and mechanical properties on the micro/nanoscales. There are many types of thin films, including self-assembled monolayers (SAM) and ionic liquids (IL). In order to decrease the frictional resistance, a hard film should be coated with soft films. Conversely, to minimize the wear particle generation, the phenomenon of plowing, and decrease the wear, a soft film should be coated with hard films. Soft films have low shear strength, low load-carrying capacity, and an extended area of contact, which leads to high friction. Conversely, hard films alone lead to high friction and high shear stress.

The failure of MEMS devices often stems from poor tribological performance and adhesion issues of the structures in the system. Thin films, which acts as the protective lubrication coatings in many applications, have poor durability because of the lack of the tribological contributions of mobile molecules, such as low shear strength, replenishment, and trapping among micro-asperities. Tailoring surface topography and chemical modification have gained momentum in improving tribological performance by reducing the area of contact between the surfaces under sliding, which changes the wetting interface behavior and traps wear debris, thereby reducing plowing effect. A dual-film coating, which includes the lubrication synergy of solid and liquid combination, has become a more popular design concept to improve the life of the lubrication coatings and thereby results in better load-carrying capacity in MEMS devices.

11.6.3 Measurement of Friction and Wear in MEMS Devices

Microsystem engineering is rich with issues concerned with the properties, processing, and mechanics of materials. For MEMS, tribology is an important enabling technology with challenges both at the stages of fabrication—especially the optimization of chemical mechanical polishing and the control of release stiction—and at device and the system operation. Much of current MEMS technology is based on silicon, and although this material has some attractive attributes-particularly its fatigue performance—it is less attractive tribologically. When running against itself, both the CoF and, most particularly, rates of wear are significantly higher than those achievable with optimized material combinations at the macro-scale. When/If the function of a machine is terminated by either the growth of excessive relative tolerances between the articulating members or by the generation of the associated wear debris, real-time life reduces as the scale falls. This means that the tribological material demands on small devices are more stringent than those of their full scale analogues, and thus, an ability to resist wear is especially cherished. In MEMS, conventional liquid lubrication is impossible because of either meniscus effects or viscous pumping or churning losses. SAMs applied to sliding surfaces may provide acceptable combinations of low friction and wear though robustness and replenishment may be issues of concern. In dry running MEMS and micro mechanical assemblies, surface treatments or alternative candidate materials must be compatible with MEMS fabrication routes. DLC has some very attractive tribological features, although some deposition techniques can result in films with unacceptably high values of residual stress.

11.7 Rail Transport Tribology

The steel wheel meeting the steel rail is fundamental to rail transport. A wheel-rail contact is an open system exposed to dirt, particles, as well as applied and natural lubrication. Given that adhesion, wear, sound, and particle emission are closely related in an open system, they should be studied together rather than independently.

11.7.1 Contact Surface of Rail and Wheel

Tribology at the wheel-rail contact is of great scientific and industrial importance because of safety and economic concerns. Tribological systems are divided into open and closed systems. If the environment is sealed, as is the case in gearboxes, closed system with a controlled environment that enables better control of friction, wear, and applied lubrication. In a closed system, a lubricant can cool the contact locations and transport away the heat generated during contact. Open systems like the wheel-rail contact or block brake-wheel contact in rail vehicles are, however, exposed to dirt, particles, and natural lubricants such as high humidity, rain, and leaves. Applied or natural lubrication as well as contaminants also affect the friction and wear processes. The tribology at the wheel-rail contact is affected by both physics and chemistry. The physical elements include surface roughness (which affects the contact mechanics) and the properties of the materials. The open system not only influences the wheel-rail contact but also receives feedback from it, for example, in the form of noise and particle emissions.

'Wheel-rail adhesion' refers to the tangential force exerted between the wheels and rails and describes the limiting friction. In railway operations, a proper CoF is desired. Contaminants are also usually present at the wheel-rail contact and affect adhesion.

Since wheel-rail contact is a rolling-sliding contact, the resulting wear can be grouped into four major classifications: rolling contact fatigue (RCF), abrasive wear, adhesive wear, and corrosive wear. Common hard solid contaminants at the wheel-rail contact include ballast, oxides, sand, and alumina, all of which have been reported to aggravate abrasive wear at the wheel-rail contact. Deep scratches, which are a sign of severe abrasive wear, have been observed on the material surfaces after testing for the presence of leaves. Snow and ice particles are another possible solid contaminant at the wheel-rail contact during winter. Pressure melting theory indicates that they will be crushed and pressed into water droplets that lubricate the wheel-rail contact, reducing the wear loss.

11.7.2 Wear of Rail Lines

The wear is caused by the contact of wheel/rail in a railway system. As we all know, the rolling of the wheel is the main operation pattern, but in traction, braking, and turning, the sliding wear corresponding to the impact between the wheel and the rail plays a significant role in the failure process of the wheel/rail system. With an increase in the sliding speed, the wear rate of wheel steel decreases and then increases. The wear pattern of some components of precision machine tools is a combination of sliding and impact wear, such as rails of machine tools, etc. Main rail wear factors include the hardness of the rail material, carrying weight, sliding speed, working conditions, and other aspects.

A lot of the complexity of the wheel-rail contact is brought about by the open nature of the system and the constantly varying environmental conditions in terms of, for instance, temperature and humidity. It is exemplified in the relationship between weather conditions and the measured rail wear shown. Here, the precipitation has a significant effect on the rail wear. Along a length of line, the position of the contact and its size and the resulting contact stresses are also continuously varying and will be different, not just for each railway vehicle, but for each wheel as each, although starting with the same profile, is worn in different amounts. Since the wheel-rail contact is an open system, damage mechanisms, such as wear and rolling contact fatigue, will be influenced by factors such as humidity and other natural contaminants.

11.7.3 Parameters on the Wheel and Rail Wear

Wheel/rail wear continues to be a very important factor limiting asset life. Wheel/rail wear simulation based on mathematical models is one of the instruments of study and prognosis of their lifespan. When simulating wheel/rail wear, it is necessary to consider the interrelation of wear process with the dynamics of the vehicle/track interaction, contact mechanical parameters, and tribological properties of interacting materials. Normal and tangential contact stresses are calculated using Hertzian or non-Hertzian Theory.

Most frequently, the wear model used in simulation models is formulated in terms of proportionality between the specific energy dissipated over the contact surface and specific mass removal for the unit distance covered. To obtain the wear model, that is the dependence of the wear rate on the contact parameters, simulation of wear between wheel flange and side face of the rail head when a vehicle moving in a curve is used. One of the main reasons that predictions of wheel/rail contact behavior are so complex is their exposure to the environment and their ability to entrain contaminants. Lubricants, predominantly grease for gauge face application in curves, are commonly used. A natural third-body layer always exists, which is an interfacial layer that is formed within the bounds of the wheel/rail contact under the high contact pressure and rolling/sliding conditions. Investigations of the constituents that make up this third-body layer suggest that they are iron oxides and wear debris from the wheel and rail. This is typical of wear by ratcheting. In practice, the stresses usually exceed the shear yield of the material, and the surface of the wheels and rails accumulate plastic deformation. The plastic flow raises the shear yield strength of the material, this process is called ratcheting. The shakedown diagram defines the limits between elastic deformation, racheting, and plastic deformation. The more the shakedown limit is exceeded, the more plastic deformation accumulates until the material's, the ductility is exhausted and failure occurs.

11.7.4 Rail Corrugation

In recent years, rail corrugation has been a serious problem. In the 1970s, a trolley-based instrument was developed for corrugation measurement as part of a long-running collaborative programme of corrugation research. Such an instrument may provide the means of obtaining accurate measurements of rail 'roughness' for acoustic purposes. The corrugation can be considered as comprising a 'wavelength-fixing mechanism' and a 'damage mechanism'. A more useful and probably less misleading classification is simply to consider the different 'damage mechanisms' that can cause corrugation, such as plastic flow, plastic bending, rolling contact fatigue, and wear. Within each of these, there may be several types of corrugation caused by different 'wavelength fixing' mechanisms, indicating different dynamic behavirs. By far, the most prevalent damage mechanism is wear, as it is associated with three of six categories of corrugation. All types of corrugation whose causes have been identified to date are associated with the resonant behavior of the vehicle/track system. They are constant frequency rather than constant wavelength phenomena. It is accordingly desirable to minimize the occurrence and effects of possibly harmful resonances. Friction modifiers offer a means of treating any type of corrugation that arises from high wear rates because of excessively high friction. Friction modifiers have been a demonstrably successful treatment of corrugation on several railway systems.

11.7.5 Wheel Wear of Metro Vehicle

Wheel polygonal wear is a typical wear phenomenon in the railway industry, which aggravates the wheel-rail interaction and prematurely invalidates or damages the components of the vehicle-track system. Since the 1990s, a lot of work has been conducted to investigate the mechanism of wheel out-of-roundness of metro trains, locomotives, and high-speed trains. With the continuous advancement of testing technology and theoretical simulation, a better understanding of the mechanism of wheel out-of-roundness becomes more possible. In terms of experiments, the wheel polygonal wear with 9 harmonics of the linear induction motor (LIM) metro train wheels is related to the first-order bending resonance of the wheelset through extensive field experiments. The P2 resonance is the reason for wheel polygonal wear with 5 – 8 harmonics of metro train wheels. The high-order polygonal wear of metro train wheels results from the first bending vibration of the wheelset, which is excited by rail corrugation with a 200 mm wavelength in the 1/3-octave band on sharply curved tracks. The first bending resonance of the wheelset contribute to the high-order wheel polygonal wear. The torsion resonance of the wheelset is the inducement of a 20th-order polygonal wear on the locomotive wheels. The root cause of polygonal wear on high-speed train wheels is the excited resonance of the bogie in high-speed operation.

11.8 Reliability of Tribology

There is an increasing need for the improved methods of determining the reliability and predicting the lifetime of machines and production systems more accurately.

11.8.1 Importantance of Reliability

One general trend in industry is to develop larger and more integrated technical systems with an increased degree of automation. This can be observed in the manufacturing industry as well as in energy production and transportation. The systems become more difficult to control, more sensitive, and vulnerable to serious consequences due to failures and breakdowns. Reliability of machines and instruments is therefore of increasing importance. Failure of technical systems may result in safety risks for people in transportation and in large environmental risks in the nuclear and process industry. Breakdowns in the production process are expensive for both the production plants and the energy suppliers. Mechanical failures of components and especially tribological failures, such as wear and friction-related failures, are today one of the main reasons for shutdowns and unavailability.

11.8.2 General Concept for Improving Reliability

There is a great variety of different techniques based on expert knowledge in several fields of technology. There is a need to approach reliability and maintainability problems from a holistic point of view, starting from the problem of the customer and ending with the satisfied user. This is aimed at improving the synergistic interactions between the different fields of expertise by showing a logical and comprehensive structure, where each expert can find their place and see the connections to experts from other fields, all working with the same aim of satisfying the end user. The critical parts are identified, the probability of system failure and lifetime are calculated, and the operability costs are estimated by statistically based techniques of reliability control. One critical part of the system that needs improvment is identified, the right techniques and tools for improvement actions are found in the fields of mechanical component failure control, electronics failure control, software failure control, or the control of human error. When a critical function is identified, such as the wear endurance life of a certain component, a component operability analysis is carried out. This includes an analysis of the old solution, a robust lifetime design approach for the recommended improvements, an analysis of the new solution, and, as a result, the improvement actions with the estimated improved failure probability and probable lifetime. The recommended measures may include changing components, implementing redundancy, improving design, extending monitoring, implementing automatic diagnosis, conducting inspections, performing operational tests, or providing service instructions. The output of the holistic approach is recommendations for improvements together with estimations of their effects on the risks, the probability of failure and the lifetime. It is possible to find the right and optimal solutions only if we manage to develop techniques and analyze the whole system as well as its components by quantitative measures. This is not, however, an easy task because of the vast complexity of all the involved reliability-related parameters and their interactions. This requirement is, though, an important challenge to all the technical and scientific fields within the broad concept of reliability.

11.8.3 Component Reliability and Endurance Life

The basic tribological phenomena friction, wear, and lubrication are all reliability-related. What is important from a reliability point of view is to produce more and better tribological data on the endurance life and on the critical wear and friction levels of tribological components, both in general and in specific operational conditions. In the first situation, severe wear begins early in the operation due to unsuccessful design. In the second situation, a mild wear condition is achieved after a running-in wear period, and a considerable safe operating time prevails until the slowly proceed wear damage of the surface results in severe wear. Finally, in the third situation, successful design has produced a contact condition, which, after a short running-in wear, results in a long period of insignificant wear. The important question from a reliability point of view is to produce the tribological data and techniques with which the endurance life can be estimated. Endurance life is the time until the component can no longer perform its planned function. The designed endurance life is the duration until the cumulative wear reaches the limit for safe operation specified in the design, the critical wear level, is exceeded in operation. Changes in contact conditions can result in exceeding of the endurance life when an increase in the friction force exceeds the values in the design specification, the critical friction level, and a risk for safe operation occurs.

11.8.4 Tribological Data for Reliability Estimation

There are four ways of getting the required tribological data for the estimation of component endurance life and risk for failure: ① by collecting historical data about how similar components have performed in similar operational situations earlier; ② by using the generic knowledge of tribological behavior in theoretical models and equations; ③ by laboratory testing of components and materials; ④ by on-line monitoring of the tribological performance. The most promising approach for collecting accurate historical data that can be used for endurance life estimation is using automatic condition and operation monitoring modules directly connected to the performing components. Then there is no human influence and all the data is collected in the same systematic way. The quantitative friction and wear modelling based on current state-of-the-art techniques has a large variation in the level of accuracy. In some areas, such as in the calculation of surface stresses or in the calculation of elastohydrodynamic contact pressures and film thickness, there are quite advanced and precise mathematically formulated models. On the other hand, there is no model available for the estimation of friction or wear in contacts with the coated surfaces in the presence of a fluid containing contamination particles. Here, the most promising approach for producing useful tribological data for reliability estimations is not to use generic wear equations but to develop component-specific friction and wear models. Laboratory testing of material combinations in specific contact situations and environments is perhaps the most reliable way of gaining basic tribological data for reliability control purposes. In addition to endurance life, another reliability parameter from a practical point of view is the probability of failure.

11.8.5 Improving Reliability by Operational Control

The task of improving operational reliability of machines can be approached from different angles. If we have earlier experience of the performance of similar machines, we can systematically collect reliability data and use this to draw conclusions for measures of improvement. This kind of information is most valuable for the design team and it is important to transfer it effectively to its knowledge base. In the development of new designs, there is not much reliable historical data available and thus the reliability estimation is more dependent on the conclusions that can be drawn based on generic knowledge or transforming information from other constructions. During operation, actions can be taken to improve reliability, with tribology playing an important role in this process. By condition monitoring of the performance, online information is obtained about the stage of functional deterioration that may result from factors such as wear of surfaces. The area of condition monitoring and diagnostic engineering is currently advanced and highly developed. It includes the development of sensors to measure changes in performance, data collection, signal processing, diagnostics, prognostics, and maintenance engineering. Other condition monitoring techniques are performance measurements, such as measuring rotational speeds, forces, displacements, temperatures, etc.

11.8.6 Tribology in Reliability-Based Design

New tools have recently been developed that take into account reliability aspects in design. The method enables the effective prediction of reliability of a product in explicit form right at the design stage and to use it. It provides a design team with the means to judge new and innovative ideas on the basis of their predicted reliability, and it helps the team to focus on the right target with their reliability engineering efforts. The method also offers the possibility to analyze the customer's view of reliability requirements and compare them with what is technically possible to achieve. The process for calculating the same reliability parameters for the product is based on the technical performance of its components and system. This is a model-based simulation approach that starts with building up a fault tree for the product. Possible failures are estimated and classified according to best, mean, and worst cases of probable occurrence. The fault tree that forms the basis for the reliability simulation is based on failure data, diagnostic and prediction models. In the case of wear or friction-related failures, the estimation of probability of appearing during certain periods in the lifetime of the component and the estimation of the probable endurance lifetime relies on the best available tribological data. This data can be obtained from historical data analysis, theoretical models, or laboratory-produced data, as described above. This emphasizes the importance for the tribology society to produce friction and wear data for different material combinations, as well as in different contact and environmental conditions in the form of endurance life or probability of failure. The data can then be utilized in reliability estimations and functional predictions for components and machines. To be of real value, the tribological data should be expressed in terms of endurance life and probability of failure. Scaling up tribological knowledge from nano- and microcontacts to solid performance estimations and reliability predictions of large machines is a major challenge facing tribological society.

11.9 Brief Summary

This chapter is about the basic components of tribology design, so journal and thrust bearing, rolling bearing, gear, mechanical seal, and wheel-rail are introduced. It also includes many theoretical analysis equations, solutions, and conclusions.

References

[1] Holmberg K, Erdemir A. Influence of tribology on global energy consumption, costs and emissions [J]. Friction, 2017, 5(3): 263-284.

[2] Maddikunta P, Pham Q, Prabadevi B, et al. Industry 5.0: A survey on enabling technologies and potential applications [J]. Journal of Industrial Information Integration, 2022, 26: 100257.

[3] Xia F, Jiang L. Bio-inspired, smart, multiscale interfacial materials [J]. Advanced materials, 2008, 20(15): 2842-2858.

[4] Gadelmawla E, Koura M, Maksoud T, et al. Roughness parameters [J]. Journal of materials processing Technology, 2002, 123(1): 133-145.

[5] Zhou W, Cao Y, Zhao H, et al. Fractal analysis on surface topography of thin films: A review [J]. Fractal and Fractional, 2022, 6(3): 135.

[6] Fox M. Chemistry and technology of lubricants [M]. New York: Springer, 2010.

[7] Kojj K. Abrasive wear of metals [J]. Tribology International, 1997, 30 (5): 333-338.

[8] Zhang J, Meng Y. Boundary lubrication by adsorption film [J]. Friction, 2015, 3: 115-147.

[9] Luo J, Liu M, Ma L. Origin of friction and the new frictionless technology-Superlubricity: Advancements and future outlook [J]. Nano Energy, 2021, 86: 106092.

[10] Leyland A, Matthews A. On the significance of the H/E ratio in wear control: a nanocomposite coating approach to optimised tribological behaviour [J]. Wear, 2000, 246(1-2): 1-11.

[11] Lu X, Khonsari M, Gelinck E. The Stribeck curve: experimental results and theoretical prediction [J]. Journal of tribology, 2006, 128(4): 789-794.

[12] Siddaiah A, Menezes P. Advances in bio-inspired tribology for engineering applications [J]. Journal of bio-and tribo-corrosion, 2016, 2: 1-19.

[13] Zhang S. Green tribology: Fundamentals and future development [J]. Friction, 2013, 1: 186-194.

[14] Bushan, Bharat, ed. Handbook of micro/nano tribology. CRC press, 2020.

[15] Miyoshi K. Solid lubricants and coatings for extreme environments: state-of-the-art survey [R]. 2007.

[16] Wood R. Marine wear and tribocorrosion [J]. Wear, 2017, 376: 893-910.

[17] Williams J A. Engineering tribology [M]. Cambridge University Press, 2005.